HUMAN RESOURCE MAN[AGEMENT IN THE]
MULTI-DIVISIONAL [COMPANY]

HUMAN RESOURCE MANAGEMENT IN THE MULTI-DIVISIONAL COMPANY

JOHN PURCELL
AND
BRUCE AHLSTRAND

OXFORD UNIVERSITY PRESS
1994

Oxford University Press, Walton Street, Oxford OX2 6DP
Oxford New York Toronto
Delhi Bombay Calcutta Madras Karachi
Kuala Lumpur Singapore Hong Kong Tokyo
Nairobi Dar es Salaam Cape Town
Melbourne Auckland Madrid
and associated companies in
Berlin Ibadan

Oxford is a trade mark of Oxford University Press

Published in the United States
by Oxford University Press Inc., New York

© John Purcell and Bruce Ahlstrand 1994

All rights reserved. No part of this publication may be reproduced,
stored in a retrieval system, or transmitted, in any form or by any means,
without the prior permission in writing of Oxford University Press.
Within the UK, exceptions are allowed in respect of any fair dealing for the
purpose of research or private study, or criticism or review, as permitted
under the Copyright, Designs and Patents Act, 1988, or in the case of
reprographic reproduction in accordance with the terms of the licences
issued by the Copyright Licensing Agency. Enquiries concerning
reproduction outside these terms and in other countries should be
sent to the Rights Department, Oxford University Press,
at the address above

This book is sold subject to the condition that it shall not, by way
of trade or otherwise, be lent, re-sold, hired out or otherwise circulated
without the publisher's prior consent in any form of binding or cover
other than that in which it is published and without a similar condition
including this condition being imposed on the subsequent purchaser

British Library Cataloguing in Publication Data
Data available

Library of Congress Cataloging in Publication Data
Purcell, John, 1945–
Human resource management in the multi-divisional company /
John Purcell and Bruce Ahlstrand.
Includes bibliographical references and index.
1. Personnel management—Great Britain.   2. Industrial relations—
Great Britain.   3. Corporate planning—Great Britain.
I. Ahlstrand, Bruce W.   II. Title.
HF5549.2.G7P87   1994   93-42896
658.3'00973—dc20

ISBN 0-19-878021-4
ISBN 0-19-878020-6 (Pbk.)

1 3 5 7 9 8 6 4 2

Set by Hope Services (Abingdon) Ltd.
Printed in Great Britain
on acid-free paper by

PREFACE AND ACKNOWLEDGEMENTS

THIS book has evolved over a period of ten years of research in large companies into the way they approach the management of employee relations and develop policies in human resource management. At the start, in the early 1980s, it was still quite rare to find academics in industrial relations devoting their attention to top management. Now it is commonplace. We felt we were charting new territory, exploring new concepts, expanding, at an alarming rate, the range of literature we needed to grapple with in order to understand the behaviour of large firms in their approach to employees. We put down markers as we went along in the form of articles, some academic, some practitioner-orientated. At times, to the confusion of our students, no sooner had we invented new categories or forms of analysis than we changed them, over anxious to try to capture what seemed the essence of what we were observing in detailed case studies, national surveys, and in hundreds of conversations with the managers themselves, in the classroom, the tutorial, or in the field.

The journey is far from ended. The pace of change in management, as in public life, has been extraordinary and unprecedented. There is a quest for big ideas, new concepts, new shibboleths. They come, often dangerously, in acronyms—TQM, PRP, QC, QWL, JIT, HRM, BS5750—and no agreement is recorded on their 'true' meaning. Flavours of the month seem to have a shorter shelf-life; there should almost be a 'use by' date enforced.

Our research and investigation continues but we felt, three or even four years ago, that it was time to try to put together something approaching a coherent view of the issues involved in human resource management in multi-divisional companies. It has taken much longer than we expected if only because other demands pressed on us—the usual case of the urgent driving out the important. We have borrowed unashamedly from the articles we published in the period of the research; thanks go to the editors of the *British Journal of Industrial Relations*, the *Journal of*

Management Studies, Personnel Review, Personnel Management, and the *Oxford Review of Economic Policy* for permission to use parts of our articles that they published at various times.

We also debated at length the style of the book. Who was it for? Were we to try to impress our academic colleagues with the robustness of our theories, the quality of our data, and our ability to make reference to every known piece of research in the area? (The doctoral student syndrome, more akin to the compulsory figure skating in ice dancing than to creativity!) In the end, since much of what we do is to work with managers, we decided to try, without compromising the need for quality research and analysis, and without slipping unduly into the classic prescriptive business-school mould of 'how to . . .', to write a practical book. If academics do not have the time to reflect on ten years of extraordinary change and try, however foolishly, to make sense of it and help the hard-pressed manager, official, or union leader, who does?

Many people have helped us on the way. The Economic and Social Research Council twice provided funds as, later, did the Leverhulme Trust in two linked projects. Doctoral students at Oxford have tested some of our earlier thinking to beyond destruction and numerous groups of students and managers have debated the ideas to our benefit. The managers in the companies covered in the research allowed us to roam where we wished and gave generously of their time. We hope we have not broken their confidence when they were prepared to speak 'off record'. The production of the book would not have been possible without the dedicated help of Ros Makris in Melbourne; Ann Bond, Lorraine Mathews, and Caroline Moody at Templeton College, Oxford, and Linda Kernoham at Trent University in Peterborough, Canada. We thank them all, as we do our publisher for his patience.

J.P.
B.A.

CONTENTS

1. INTRODUCTION 1
   The research programme 4
   The structure of the book 8

2. THE NATURE AND DEVELOPMENT OF THE MULTI-DIVISIONAL COMPANY 11
   Diversified, multi-divisional companies 11
   Conclusion 26

3. THE STRATEGY CONCEPT 27
   Strategic human resource management 29
   Links between corporate and HRM strategy 32
   The strategic choice perspective 37
   Three levels of strategy 42
   Summary 48

4. CORPORATE STRATEGY AND THE INFLUENCE OF PERSONNEL 50
   Types of strategies 50
   The influence of human resource issues on corporate strategy 55
   The influence of corporate strategy in human resource management 62
   Critical corporate human resource management decisions 64
   The effect of financial control systems 75
   Summary 77

5. THE CORPORATE PERSONNEL DEPARTMENT 82
   Organizing the personnel department in the multi-divisional company 82
   How big should the corporate personnel department be? 87

Determining the corporate personnel role	100
Nine roles for corporate personnel departments	107
Summary	115

6. STRATEGY AND THE STRUCTURE OF COLLECTIVE BARGAINING — 118

Some key definitions — 118
The evolving structure of collective bargaining in the UK — 120
Forces encouraging bargaining structure change — 129
Strategic choice in collective bargaining: centralized v. decentralized — 134
Advantages of different bargaining structures — 136
Choosing the best bargaining structure — 143
The politics of bargaining decentralization — 151
Summary — 162

7. MANAGEMENT STYLE IN THE MULTI-DIVISIONAL COMPANY — 165

The importance of management style — 165
The management style matrix — 176
– individualism — 179
– collectivism — 182
– linking individualism with collectivism — 184
– the components of management style — 188
Changing management style — 201
Summary: Management style as a strategic tool — 211

Appendix 1: Rover Tomorrow — 216
Appendix 2: Note on the Main Statistical sources used — 219

References — 221

Index — 231

1
INTRODUCTION

Take any textbook on corporate strategy and business policy and look up anything relating to employment in the index. The chances are that no reference will be made to staff, workers, employees, trade unions, personnel, human resource management, or industrial relations. Now take personnel management or industrial relations textbooks and try to find references to business policy, corporate strategy, or specific techniques such as portfolio planning. There might be some references, but a reading of the text will reveal only a vague understanding of the impact of 'strategy' on personnel management and a common injunction that businesses ought to include human resources policies as part of their corporate strategy. Sometimes it might be argued that personnel policies are a vital component in corporate strategy, but the implication is that, all too frequently, they are not regarded as such.

It is as though there are two separate worlds of analysis and action. Business policy ignores the human dimension; specialist personnel or human resources management strategies are pursued, it seems, with scant regard to the wider business context. This is not simply a problem for academics and students but reflects the bifurcation of business practice. It is relatively easy to see the development of employee relations strategies concerning for example employee involvement, union recognition, payment systems, and recruitment practices, but hard to see the connection with business policy except in the vague sense of responding to competitive pressure or a requirement for flexibility. It is the central thesis of this book that strategies in employee relations can only be examined in the context of the business in which they operate, and that long-term business policy needs to take account of the choices and constraints in employee relations strategy within and outside the firm.

While it is generally accepted that employee relations strategy and planning tends to have little influence on corporate strategy (Winkler 1974; Hickson and Mallory 1981) and that the development of employee relations policies will be deeply influenced by the wider corporate strategies that the enterprise adopts (Kochan, Katz, and McKersie 1986), little discussion has taken place about the precise nature of these links. Most discussion to date has been developed at the general level and has not differentiated between different types of company strategies and structures and attendant links into employee relations practice. In this book we focus principally on strategic considerations in diversified or multi-divisional companies.

The justification for focusing on multi-divisional companies is threefold. First, evidence reveals that many advanced industrial countries are dominated by large diversified companies and that mergers and acquisitions are increasing the number of establishments they own and often broadening their range of activities. Large single-product firms are declining, while related-business and conglomerate-business firms are growing. Secondly, by virtue of their size, the mixed portfolio of businesses owned, and by their dominant position in certain markets, these firms have a greater degree of strategic choice as to how to respond to and influence their environment than traditional, functionally organized firms. To a degree the invisible hand of the market has been replaced by the visible hand of the corporate office.

Thirdly, such firms are increasingly adopting the multi-divisional form of organization and there is strong American, and some British, evidence that these firms exhibit a superior economic performance. Briefly put (much more detail will be provided in the next chapter), the multi-divisional company is structured to maximize its exposure to the various product markets it serves, whether these are defined in a product or geographical sense. Businesses are organized into divisions and/or business units with business unit managers made wholly or largely responsible for operating decisions and for profit generation. Head offices are responsible for broad strategic decisions and, critically, develop and manage the internal capital market such that all profits generated by business units are the property of the centre to dispense in the light of strategic need. Financial or performance-control systems, usually seen in annual budgeting and monthly reporting, are used to monitor and motivate business unit performance. It is argued that it is impossible to understand human resource management in these firms without appreciating the way multi-divisional companies are structured and behave. Research data tells us that companies are increasingly turning to multi-divisional forms of structure (Channon 1982). Unfortunately, however, this revolution in restructuring has not been matched by a growth in ways of understanding the dynamics of this organizational species. In particular, we know little about the way in which people (or human resources) are managed within the context of this particular organizational structure.

It is possible to argue, in fact, that the bulk of the personnel and human resource management literature has been blind to this new wave of organizational restructuring. Most literature tends to be based around 'functionally' organized companies and, as a result, bypasses the consideration of how personnel management is carried out in other organizational configurations. The implication of this is that students of personnel and human resource management are using outdated and redundant concepts and theories which are more applicable to old organizational structures. The literature has provided little (if any) guidance on the unique problems associated with managing personnel in multi-divisional structures. In short, there is a serious mismatch between the maps provided and the organizational terrain that students of personnel enter.

The mainstream personnel and human resource literature has simply failed to differentiate between organizational structures and has, rather naïvely, assumed that the same human resources issues that apply to one organizational structure will apply to all. This lack of differentiation has reached the extreme when researchers studying the human resources of particular operating units fail to recognize (and fail to incorporate into their analysis) that these units are often part of a much larger and complex organizational structure. Many of the more significant human resources issues may actually be determined outside of these plants, in the higher reaches of the corporation. The artificial boundaries placed by researchers around the workplace have served to obscure the distinctive and complex features of human resource management within the multi-divisional company (Purcell 1983). The independent, functionally organized company of the 1960s, captured to a degree by its environment, size, and technology is now, in all probability, the subsidiary company or business unit of the multi-divisional corporation.

It seems sensible, therefore, to ask how the behaviour of multi-divisional firms influences the management of people at work, through the business strategies that these firms adopt, the strategic choices that they make within the field of human resource management, and the controls that they exercise on the behaviour of business units and subsidiary companies.

THE RESEARCH PROGRAMME

This book derives from a variety of research projects undertaken over a ten-year period. In one sense it can be seen as a product of continuous 'action'-based research, extending over the whole of this time-frame, but especially the five years from 1982 to 1987 when research funding was received from the ESRC. As well as relying on our systematic case-study work, the research taps into our own experiences and encounters with a variety of practising managers in different settings. In addition, we look to those few researchers who have conducted work in this area, in particular the Warwick research team and the results of their 1985 workplace industrial relations company-level survey (Marginson et al. 1988) and the subsequent 1992 company-level survey (Marginson et al. 1993). Much of what we have to say in this book is derived from our case-study research, which was undertaken in two periods.

In 1983-4 two studies were completed in 'foods' and 'leisure' (for more details see Purcell and Gray (1986)). In 1985-8 a further seven studies were undertaken in large enterprises. Some were chosen because their centralized system was known to be under review (privatized corporation, bank, transport services), others because of their strong emphasis on divisionalization (motor components) or decentralization (staple foods). One firm was mainly non-union (electronics), while the last was the major division of a retail firm running department stores in London (retail). All of these enterprises were structured along divisional lines, although the size of the major division constituted over 80 per cent of the employees in two cases. A full list of the companies is provided in Table 1.1.

TABLE 1.1. *The panel of companies*

Company pseudonym	Main research period	Further data collection 1990–3	Size range: employees (UK)	No. of main UK divisions at time of study	No. of interviewees	Restructuring of company in period of research	Collective bargaining	Internal labour market (job evaluation)	Main board Personnel Director 1984	Main board Personnel Director 1992
1. Foods	1983–4	✓	25,000+	6	45	Yes	Divisional but shifting to plant	Divisional	✓	×
2. Leisure	1983–4	✓	23,000	5	40	Yes	Division/Local	Local	✓	×
3. Bank	1985–6		40,000	5	64	Yes	Multi-employer to corporate level	Centralized	×	×
4. Staple	1986–7	✓	18,000	6	35	Yes	Multi-employer plus highly decentralized	Highly centralized	✓	×
5. Retail	1986–7		10,000	3	40	Yes	Some divisional and local store level or none at all	Local store and floor level	×	×
6. Privatised corporation	1986–7	✓	200,000+	5	35	Yes	Corporate	Centralized	×	×
7. Motor components	1986–7	✓	20,000+	4	15	Yes	Corporate/Regional to plant	Becoming decentralized	✓	✓
8. Transport services	1986–7	✓	8,000	3	32	Yes	Corporate to two-tier	Corporate	×	×
9. Electronics	1987–8		7,000	2	10	Yes	Union recognition withdrawn No bargaining		×	×

In each case extensive interviewing was done at corporate, divisional, and operating unit levels of personnel and industrial relations executives and some line managers and corporate executives. The central focus was strategic choices in employee relations (defined, loosely, as a combination of personnel and industrial relations) and an exploration of the factors within the enterprise which shaped and helped to explain the strategic choices taken, or not taken, within employee relations. Interviews were conducted with a selection of managers at all levels of the enterprise to explore dimensions of the 'double bind' which our pilot survey revealed to be prevalent, namely the expected tension between strategic and operational management (how far monitoring of plant activities led to intervention, for example), and between personnel 'professionals' and line managers in policy formulation and implementation. A key feature of the research was the focus on 'key incidents' in the internal environment (e.g. how do decisions on take-overs or rationalization influence the management of employee relations?) and external environment (how does the corporation respond to changes in product markets or the law?), designed to elucidate actual behaviour in context.

Gaining access to do research in an M-form company is complicated by the fact that approval often has to be negotiated at the three separate levels of the central office, division, and plant. Obtaining approval from central office to proceed with the research is not always sufficient in itself. Even in those M-form companies that are centralized, the central office is often wary about giving approval to proceed with the research at lower levels as it often wants to maintain some impression or illusion of autonomy of decision-making at lower levels in the organization. On the other hand, highly decentralized companies, often as a matter of policy, would not permit their central office to make a decision on research access for their 'autonomous' business units. The need for the researcher to deal with three separate levels of the organization on access matters makes the process perilous and time-consuming.

There is also a problem with sorting out a host of 'discrepancies' between attitudes and actions at various levels in the organization. Much of our own research was spent attempting to sort out differing views between levels of the organization: views on what is the formally espoused strategy and the informally 'accepted' policy. These views have sometimes been found to differ so markedly, between, say, the centre and the plant, that the researcher can often wonder if s/he is, in fact, in the same organization. In the privatized corporation we were asked whether we wanted to know about 'stated policy or done policy'. The answer, of course, was both.

One example will suffice to show the complexity of the research problem. In 'staple products', a highly decentralized firm, we asked the corporate personnel director if guidance was given to operating companies and plants on collective bargaining pay offers and settlements. The answer was an unambiguous 'no, we leave that to the negotiators. They know what is best for their business'. We asked the same question at the divisional level. Here we were told that 'we hear what the centre has to say since there is always someone from Group on the divisional Board'. At the plant level the managers who undertook the negotiations with the unions were equally clear. 'Yes, of course there is a mandate and I am expected to discuss our negotiations with my opposite number in other divisions and companies. We also tell head office.' We went back to the Group corporate office in London: how could this discrepancy be explained? 'Well, I suppose it's what we call "management by raised eyebrow". They just get to find out what is acceptable and what isn't!'

This book, as suggested, also draws on our experiences with practising managers. Over the period we were drawn into association with managers in a variety of settings and for a variety of reasons. As the research progressed we became involved (sometimes unwittingly) in a number of different consultancy activities. We interpreted the relatively large number of requests for consultancy work, however, as an indication of the relevance of our research programme. Senior managers of large multi-divisional companies were confronted with a number of complex human resource problems and there was simply little academic research which could help them resolve them. Many questions were posed. Many of these related to the 'bargaining decentralization' issue (should we decentralize our bargaining? if so, to what level, the division or the plant? how should we do it? what problems might we face if we embark on this change?); others related to the role of the corporate personnel department itself (what should central personnel really be doing? should we devolve our personnel activities downwards to division or plant level? if so, how can we do it and what are the potential repercussions here?).

Needless to say, the concepts, models, and theories which we brought to bear on these problems were in many cases hugely inadequate, but as our involvement in these consultancy activities grew, so did our sophistication in dealing with these problems. While we will not comment on the efficacy of our own consultancy, we can only be thankful to those organizations for providing a laboratory setting for the understanding of human resource management issues in the multi-divisional setting. Our exposure to practising managers did not begin and end with our consultancy. The lessons of this book are just as much a product of our exposure to them in these training and consultancy activities as of our more systematic case-study work. In many cases it was the practising managers who posed and raised the important problems and provided useful insights into how they managed themselves out of these problems.

THE STRUCTURE OF THE BOOK

Chapter 2 begins with a definition of terms. First we attempt to define for the reader what we mean by a multi-divisional firm. The reader will see that the definition is an elusive one. We also provide for the reader a sense of the growth in incidence of this particular organizational structure. Chapter 3 discusses and elaborates upon the 'strategy' notion. Here, we hope, we provide some new ways of looking at strategy in each of the corporate, human resource management, and multi-divisional contexts. Here we also look at the links between corporate and human resource management strategies (as they have been articulated in the recent literature).

Chapter 4 builds on our conceptualization of strategy and develops a model of strategic decision-making and influence which can be applied to multi-divisional companies. This conceptualization posits a three-level model: first-order strategic decisions (long-term direction of the firm), second-order decisions (internal operating procedures) and third order (strategic choice in employee relations). Once the model has been elaborated, both the influence of human resource strategies on corporate strategies and the influence of corporate strategies upon employee relations strategies are considered. Special emphasis is placed on diversification strategies. The chapter argues that diversification strategies tend to lead to specific patterns of human resource practice.

Chapters 5, 6, and 7 are devoted to an investigation of the strategic process within the employee relations sphere. The organizational structure of the M-form company itself, consisting of corporate headquarters, division, and operating subsidiaries, defines, by its nature, a rather unique set of strategic choices in human resource management. The existence of various levels within the organization and the existence of various business units means that the organization has, for instance, a choice about: (1) the level at which it wishes to conduct its collective bargaining; (2) whether it wants to segment its labour market by division or unit; (3) whether it wants to create a single dominant corporate culture or, at least, allow the existence of a competing set of cultures or styles; and (4) what kind of division of personnel responsibility should take place between the various levels of the organization. Evidence from our own case studies reveals that companies are, in fact, acting strategically, that is they are either reviewing or are in the process of making fundamental choices or decisions about their human resource management system or structure (although the attempt to make change has not always been met with the desired results). Choices that are made (or not made) across these four decision areas set the tone and determine the level of responsibility for managing employee relations. These key areas of strategic choice are evaluated within a context of evolving organizational structure, including the development and break-up of divisions, mergers, acquisitions, divestment, and the use of profit centres and internal control mechanisms.

Most research on employee relations management in the multi-divisional company has focused on some specialized area of strategic choice; Kinnie (1987), for example, on choice of bargaining level and Purcell and Gray (1986) on the division of responsibility within the personnel function. The implicit assumption of such research has been that strategic choices taken within the employee relations system are somehow disconnected from one another. This book, on the other hand, takes an integrated approach to strategic choices within the multi-divisional company and places stress on the links between choices within the employee relations system.

There is, for instance, a need for a certain coherence in the choices taken within the human resource management system as an adjustment to changes in other parts of strategic management. Decentralizing or devolving the bargaining structure from corporate to plant level, for example, will clearly have implications for the division of responsibility within the personnel function and most notably for the role of the corporate personnel department. Changing the bargaining level will also have important implications for leadership style and culture. Exercising choice within the human resource management system is, of course, also conditioned and sometimes constrained by a set of wider 'first-order' corporate or business policy choices and internal control mechanisms. Our research evidence suggests that strategic choice around the four fundamental decision areas is ultimately determined by 'higher order' corporate strategy and business policy.

This book represents the first systematic attempt to uncover and analyse the way in which human resources are managed within multi-divisional structures. We hope that it will stand as a map for students to use both to understand the complexity of human resource management in the multi-divisional company and to provide a basis for action.

2
The Nature and Development of the Multi-Divisional Company

The first part of this chapter sets out what is meant by a multi-divisional company. The second part charts the development and evolution of this organizational form. The final section seeks to provide different explanations and theories about its proliferation.

DIVERSIFIED, MULTI-DIVISIONAL COMPANIES

Asked to draw an organizational chart of a typical company, most students produce a hierarchical, functional model. The board of directors, headed by a managing director, chairman, or chief executive officer, sits at the apex surrounded by functional managers for production, marketing, sales, finance, and personnel. Some may show dotted, non-reporting lines between functional areas and others may be more precise about board membership, or sub-board executive committees where the functional heads meet. This is often an adequate description of small to medium-sized single-product companies, and can still be found in some large, highly centralized single-business corporations. The assumption is that the firm is an independent, stand-alone unit operating within the confines of the external product and capital markets. The distinctive feature of the functional organization is that people and activities are grouped together by resources. All similar and related occupational specialisms are grouped together. Each functional department provides resources to the overall production process of the organization. This form evolved at the turn of the century, partly in response to the increasing size and complexity of business organizations. It is also referred to as the 'unitary' or U-form company. The term 'functional' comes from the idea that there are functions or activities that the organization must perform if it is to carry on its business (Jackson 1986: 167). It is this structure on which most of the personnel/human resources literature is based.

The multi-divisional company is, however, a beast of quite a different nature. In this structure, each unit or division is relatively self-contained in that it has the resources to operate independently of other divisions. Each division is typically headed by an executive who is responsible for investment in facilities, capital, and people as well as for the division's development and performance. This structure is similar to dividing an organization into several smaller companies, with the chief difference being that each 'smaller' company is not itself completely independent. The most basic feature of the divisional structure, therefore, is 'self-containment'—each division is like a self-contained organization. This means that each division will have all of its own occupational specialisms, in order to develop, produce and market its particular array of products. In fact, once a company has 'divisionalised', divisions, or the operating companies within divisions, are typically organized along functional lines again. In a divisional structure, divisions can be organized according to individual product groups, services, regions, markets, customers or major programmes. By far the most popular M-form configuration is the product structure. The distinctive feature of a product structure is that grouping is based on organizational outputs.

M-form companies differ widely in their operating arrangements, but typically exhibit a three-tiered structure: a corporate central office, divisions, and operating plants or establishments (within the divisional structure itself). Typically, the corporate office provides assistance to the divisions, overseeing, controlling and co-ordinating them. The division manager, while having responsibility and authority for the performance of the division, is usually subject to some degree of evaluation and control by the corporate office itself. A key issue here is precisely what the division managers can decide on their own and how far they can go independent of the firm's policies. Not surprisingly, the degree of divisional autonomy varies considerably.

In multi-divisional companies there is, in theory at least, a separation of profit responsibility at division level from strategic management at corporate level. The term 'division' is used to cover a

Development of the M-form Company

number of different types of organizational structure, based on market or regional divisions. Operating companies and/or strategic business units are covered in the generic term. Actual organization within the multi-divisional companies can be highly complex with, for example, profit centres forming accounting units or operating companies, which are themselves grouped into strategic business units reporting to divisional structures, which in turn are owned by the corporation. Thus, divisions are responsible for the production and marketing of a major product market' while 'the corporate office . . . besides coordinating, monitoring and evaluating divisional performance . . . plans for the continuing health of the enterprise and allocates the funds, equipment and personnel required to carry out such plans' (Chandler 1976: 23).

Three features of this structure are important for our purposes. First, as Chandler points out, in such firms corporate headquarters is responsible for business policy and planning, defined as the identification of strategic long-term goals, the development of guiding policies and courses of action, and the allocation of resources necessary for goal attainment. These planning activities might include market research and portfolio development (entering growth areas and exiting from mature or declining markets) and using analysis of internal strengths and weaknesses to identify divisional investment priorities, approving budgets, and establishing targets for financial and market performance. In certain circumstances more emphasis might be given to one of these functions than to the others. Secondly, profits are not automatically returned to the divisions which generated them, but are instead subjected to a system of divisional bidding, with corporate management deciding on the allocation between competing demands in the light of strategic need, usually determined by an assessment of market growth potential, market share, and competitor analysis. Thirdly, an extensive set of controls is established which seeks to regulate and monitor divisional behaviour. As Williamson and Bhargava (1972) note, formal controls over divisional behaviour and performance include regular financial audits (or reporting requirements), while more informal controls include the gain or loss of professional status and promotion. Taken together, these controls provide corporate management with comprehensive data on divisional performance and tie both managerial rewards and future funding to business success. Divisional freedom of operation exists only as long as divisions or their sub-units continue to fulfil corporate needs and expectations. Those who fail to attain the desired market share or return on investment for example, or those who are caught in declining markets, often find themselves the subject of reorganization, or fall victim to corporate intervention or divestment. Thus, not infrequently, the decentralized or devolved organizational structure typical of M-form companies actually results in the centre retaining considerable control.

A major distinction between M-form companies and their more traditional counterparts is the use of budgetary control systems with monthly reporting and the lack of hierarchies linking functional heads in divisions to corporate headquarters. Most corporate boards are composed of divisional managing directors, the chief executive, and the finance director (the key functional specialism) with outside non-executive directors.

The development of internal capital markets within multi-divisional companies also gives substantial choice to corporate executives as to how they expose subsidiary companies to capital demands. The ability to switch production or operations from one site to another (and often between countries) and the very size of the corporation provide considerable power in the labour market. The diversified nature of the firm and its portfolio of businesses allows it to exit and enter certain markets with relative ease. The ability to dominate some product areas allows it to escape the vagaries of the product market and, to a degree, to determine price structures in an oligopolistic or monopolistic way. The economic power of these companies is associated with political and social power such that business leaders are, to a degree, able to influence the external environment by supporting political parties, lobbying federal, central, and local governments, and courting political favour through investment strategies. The very largest companies have sales revenues in excess of the GNP of many nation-states in Europe as well as the Third World (Morgan 1986: 300). In short, they are capable of exercising strategic choices over a much greater range of issues and with considerably more discretion and power than traditional firms.

In their simplest and purest form then, M-form companies consist of a number of semi-autonomous operating divisions, business units, or wholly owned operating companies, established on either a product or a geographical basis. While it is easy to talk in terms of 'pure' M-form structures, in reality we see a great variety, many of which could be defined as 'hybrid' structures. Most corporations, in fact, do not have a pure functional structure. As Daft (1989) suggests, when a corporation grows large and has several products or markets, it is typically organized into self-contained units of some type. Functions that are important to each product or market are decentralized to the self-contained units. Other functions are centralized and located at corporate headquarters. Headquarters functions are relatively stable and require economies of scale and in-depth specialization. By combining characteristics of both functional and product structures, corporations can take advantage of both forms of structure and avoid some of the weaknesses of each.

An important development of the divisional product structure has been the strategic business unit (SBU). SBU structure is similar to product structure, because it brings required functions together as needed to produce organizational outputs (Daft 1989: 232). It is now also commonplace to talk in terms of 'conglomerate' structures. A conglomerate is an organization engaged in multiple sets of different businesses. In the conglomerate structure there is very little interdependence of businesses, with the exception of the pooling of financial resources. Conglomerates resemble 'divisional' structures, but differ in terms of the relationship of business units with the centre. The distinction between conglomerates and divisional structures is often one of degree. Large conglomerates typically use a divisional and/or SBU structure in which all activities associated with major divisions or subsidiaries exist under a specific general manager, and business strategy, including acquisitions for example, is placed in the hands of the divisional board.

While we are pointing to the need to differentiate strategy by organizational form (between, say, M-form and U-form companies), we are equally sensitive to the arrangements and operating procedures within particular organizational forms. Our own preliminary case study research has revealed rather distinct differences in organization between different M-form companies, in degree of centralization and in control systems and apparatus, so that the notion of the M-form company exists more as an ideal rather than a real type. Some of our case-study companies had very complex divisional structures (divisions within divisions) while others had no divisions but operated within a complex plant or subsidiary unit-based structure. Another of our case-study companies was organized functionally with a 'dominant division' which operated alongside a number of smaller subsidiary companies. In this case, the dominant business developed into an M-form configuration through a process of acquisition of related businesses. Clearly there is a wide variety of shapes and structural forms that an M-form company can take, but in each to a greater or lesser extent the development of the M-form structure and process is associated with what we call the three Ds: diversification, divisionalization, and decentralization. There is also a fourth D, divestment, as enterprises move out of one activity into another, or return to their core businesses, as has been the case in the early 1990s.

An investigation of employee relations strategy in the M-form company will need, therefore, to be sensitive to the considerable variation in operating procedures between M-form companies and will need to develop models of strategic decision-making which take into account these variations. The link between these different organizational forms however, is that all multi-divisional companies, whether nationally or multinationally based, have a theoretical capacity to exercise choice in the way that they seek to, and actually do, manage labour relations and wider human relations questions and to link these choices to business strategy.

M-form companies have grown like Topsy and have come to dominate the economies of both sides of the Atlantic. It could be said, in fact, that the shift from functional to M-form structures has represented the single most important trend in organizational restructuring in the last fifty years. It is this organizational form which is increasingly used as the model for nationalized industries, utilities, and the public service sector; research on trends in restructuring is unequivocal. In the USA, for example, in 1949 approximately 63 per cent of the largest 500 firms were organized along functional lines. By 1969 only about 11 per cent used this form (Rumelt, 1974). At least, for the largest firms, the product division form of organization is now the dominant form of organizational structure. Rumelt notes for example that, as early as 1969, about 76 per cent of the 500 largest firms in the United

States were organized on a product division basis (only 1.5 per cent used the geographic division form).

The picture for Europe is much the same. Franco (1974) notes for example that, prior to about 1960, most European firms employed the basic functional pattern. By about 1972, however, about 70 per cent of the firms he studied had transformed themselves into the divisional structure. While many of these firms had not fully adopted the American practice of the central corporate office managing total organizational performance, the same divisional pattern employed in the USA was employed for the most part in Europe.

The British case is particularly interesting. What we are witnessing here is both a growth in the size of corporations and an associated move towards M-form organizational structures in both the private and the public sectors. Estimates of the size of diversified companies are surprisingly difficult to obtain, especially in the service sector. Table 2.1 focuses on the British manufacturing sector. In 1989 half of the employed workforce in the sector worked for 587 enterprises, each with 1,000 or more employees; a third worked for 116 enterprises with 5,000 or more people and 22 per cent in the 47 firms with 10,000 or more employees. These enterprises were even more important in the sector when their

TABLE 2.1. *Enterprise size in UK manufacturing, 1989*

Enterprise size (no. of employees)	1,000+	2,000+	5,000+	10,000+	20,000+
No. of enterprises	587	293	116	47	18
No. of establishments owned*	6,601	5,129	3,257	1,895	1,081
Av. no. of establishments/businesses owned	11	18	28	40	60
Av. no. of employees per business/establishment	380	409	480	566	633
% of employees in manufacturing in these enterprises	52	43	32	22	14
% of wages and salaries in manufacturing	56	47	36	24	15
% of gross output in manufacturing	60	52	39	26	17

*Excludes businesses employing fewer than 20 persons.
Source: Report of the 1989 Census of Production, PA 1002 (HMSO, 1991), table 12.

TABLE 2.2. *Average number of establishments/businesses owned by large enterprises in UK manufacturing, 1958–89*

Enterprise size (no. of employees)	2,000+	5,000+	10,000+	20,000+
1958	12	21	30	44
1978	19	29	42	56
1985	20	37	40	60
1989	18	28	40	60

Sources:
1958: Historical Record of Census of Production, 1907–1970.
1978: Report of the 1978 Census of Production, PA 1002 (HMSO, 1981), table 12.
1985: Report of the 1985 Census of Production, PA 1002 (HMSO, 1988), table 12.
1989: Report of the 1989 Census of Production, PA 1002 (HMSO, 1992), table 12.

contribution to gross output was examined, and especially when one looked at the proportion of the total wages and salaries bill in manufacturing. The multi-establishment nature of these large firms is clearly shown in the table, as is the fact that, on average, the number of employees per establishment increases as enterprises grow in size.

In the late 1970s, before the major recession of the early 1980s, there were more large firms in manufacturing and their share of employment was even greater. The substantial decline in industrial employment from 44.8 per cent of the labour force in 1970 to 28.8 per cent in 1990 (Crafts 1991: 85), was especially marked in large enterprises and large establishments. Table 2.2 shows, however, that the average number of establishments owned by large firms continued to increase until the mid-1980s so that, while numbers employed have fallen, the accumulation of establishments owned by large enterprises has continued. This is probably a reflection of the increasing pace of mergers and acquisitions and the investment in greenfield sites. The Census of Production now refers to establishments as businesses in recognition of the diversification of business portfolios.

The concentration of employment in manufacturing is not the whole story. Table 2.3 shows the changing sectoral composition of the workforce in Britain with projections forwards to 2000. The decline in manufacturing from 35 per cent in 1954 to an

TABLE 2.3. *Employment by broad industrial sector in the UK, 1954–2000*

	Share total employment (%)					
	1954	1971	1981	1991	1995	2000
Primary and utilities	10.7	6.3	5.5	3.8	3.3	2.9
Manufacturing	34.6	33.5	26.2	19.7	18.7	16.9
Construction	6.1	6.3	6.3	6.4	6.2	6.2
Distribution, transport, etc.	24.8	24.7	26.4	27.2	27.0	26.8
Business and miscellaneous services	9.0	11.7	15.4	22.1	23.2	25.3
All industries above	85.2	82.4	79.6	79.3	78.5	78.0
Non-marketed services	14.8	17.6	20.4	20.7	21.5	22.0
Whole economy	100.0	100.0	100.0	100.0	100.0	100.0

Note: Employment levels relate to annual average figures covering employees in employment and the self-employed.
Source: Institute for Employment Research (1993).

expected share of 19 per cent by 1995 and 17 per cent at the turn of the century is especially marked. It would not be surprising to find that firms in the manufacturing sector were seeking to diversify into the services sector.

A similar concentration of economic and employment power is found in the USA, where in 1983 the top 100 manufacturing companies controlled 48 per cent of the total assets in the sector. In Australia in the late 1980s the top 100 companies employed approximately one and a quarter million people which, as in the UK, was equivalent to just under a quarter of the working population in employment. The employment concentration of the top ten Australian companies was even more marked than in the UK or USA, with a workforce equivalent to over 10 per cent of the employed population. In terms of workplaces or establishments, 1 per cent employed 500 or more employees, but 24 per cent of employees worked in these establishments (Callus 1991: 19). Of even greater interest is the fact that 95 per cent or so of all workplaces employing 200 or more employees were owned by and part of larger organizations. This means that two-thirds of Australians in employment work in establishments owned by larger organizations. Many of these will be organized along multi-divisional lines.

Recent figures for the European Community from the University of Warwick provide comparative European data

(Table 2.4). Non EC, foreign-owned firms are excluded. The dominance of big firms in the UK is especially noteworthy, as is the size of the non-manufacturing firms, although some of these are classified as holding (or more often, conglomerate) companies with substantial activities in both camps.

TABLE 2.4. *Undertakings in the EC with 1,000+ employees (number by size of group)*

Size Group	1,000–4,999		5,000–9,999		10,000–19,999		20,000+		TOTAL
	M	NM	M	NM	M	NM	N	NM	
Belgium	77	85	6	6	5	5	0	3	187
Denmark	48	38	4	5	2	3	0	0	100
France	321	405	27	49	13	23	6	29	873
Germany	1,107	939	109	107	40	63	36	48	2,449
Greece	32	11	0	2	0	1	0	0	46
Ireland	20	27	0	13	0	5	1	1	67
Italy	259	147	25	18	4	15	5	6	479
Lux.	3	5	0	0	0	1	0	0	9
Neth.	102	475	8	49	3	28	6	29	700
Portugal	73	68	5	10	0	3	0	3	162
Spain	143	144	18	24	8	5	2	7	351
UK	680	1,582	62	276	38	161	20	205	3,024
TOTAL	2,865	3,926	264	559	113	313	76	331	8,447

M = Manufacturing
NM = Non-manufacturing
Source: Sisson *et al.* (1992: 8).

The *Financial Times* 500 list of the top European companies is especially revealing, with 72 of the largest 200 companies (by market capitalization) being registered in Britain in 1992. In total they employed 3.35 million people. Of these, eleven were mainly in manufacturing, thirteen in banking and insurance, seven in retail, eleven in utilities, eleven in food, drink, and tobacco, eight in leisure and health care, and four in transport. Three companies (BAT, Hanson, and Inchcape) were described as diversified holding companies. In practice most of these companies are diversified beyond a single industry and the stock market classification is misleading. Take P & O Steam Navigation for example. Not surprisingly this is listed as 'sea transport', but in practice it owns twenty shipping companies, eleven construction and house-building firms, six property firms, and fifteen service businesses. Sea transport only contributed 38 per cent to turnover in 1991.

Development of the M-form Company

Not all of these large enterprises are diversified, and not all of them have a multi-divisional structure, but the evidence is clear that diversification and divisionalization have been growing rapidly and are likely to continue. Prais (1976: 20) observed that in 1958 there were sixteen large enterprises in the UK active in ten or more industrial sectors. By 1963 the number had doubled to thirty-two. At the same time the number of large enterprises whose activities were limited to a single industry fell from thirty-eight to nineteen. Channon (1982) estimated that by 1980 only 9 per cent of the top 200 UK companies were 'single businesses', where 95 per cent of sales came from the one business. In 1950 the comparable figure was 35 per cent. A quarter of enterprises in 1980 were 'dominant businesses', with 70 per cent to 95 per cent of sales coming from a single business. Just under half operated in 'related businesses', where less than 70 per cent of sales came from any single business, but relationships existed between the businesses owned. Finally, 18 per cent were classified as 'conglomerate businesses', where sales were distributed amongst a series of unrelated businesses such that no one business accounted for 70 per cent or more of sales. In 1950, 38 per cent of the top 200 companies were 'single' or 'dominant businesses'; by 1980 this had dropped to 17.5 per cent. The biggest shift was in 'related businesses', which grew proportionately from 10 to 24.5 per cent (ibid.). Since the collection of these data the substantial merger and acquisition wave of the mid-1980s is likely to have altered the picture with a probable further growth in related and especially conglomerate businesses, seen in the rapid emergence of firms like Hanson, BTR, and Williams Holdings.

The more diversified the enterprise the more likely it is that the internal operating procedures will change and a form of multi-divisional structure will be adopted. Functionally organized firms are inevitably centralized and experience difficulty in devolving decision-making in increasingly diverse and complex businesses. One solution has been the development of the holding company, where the centre plays the role of friendly banker leaving wholly-owned businesses free to determine their own strategy, retain profits, and control investments and performance. This loose relationship shelters subsidiaries from the rigours of the capital market and, if supported in loss-making, from the consequences of poor performance in the product market. It does little, however, to encourage improved performance, monitoring and control prove difficult, and the reallocation of resources between subsidiaries is hard to achieve.

The multi-divisional firm emerged in the USA as the most effective way of managing diversified business, combining tight control over the financial performance of subsidiaries, restructuring of the firm to meet product market needs through the creation of product market related divisions, and the centralization of resource allocation and strategic planning. The multi-divisional structure is commonly seen as 'a superior organizational form since it separates strategic from operating decisions. It makes visible the contribution of each division to profitability and hence increases the probability of optimising resource allocation within the organization' (Hill and Pickering 1986). The authors cite a variety of studies, all finding that multi-divisionals are generally associated with superior profitability in the USA. While the picture is more mixed in the UK, studies by Steer and Cable (1978) and Thompson (1981) give evidence of better performance in profit terms.

There appear to be quite a wide and complex range of motivations for the movement from functional to M-form companies. A number of researchers (Allen 1958; Grinyer *et al.* 1980) have explained this shift in terms of an increase in desire to pursue a 'diversification' strategy. Grinyer *et al.* analysed the relationship between corporate strategy, organizational structure, perceptions of the environment, and the financial performance of forty-eight companies in the United Kingdom. Their study found that more diversified firms employed divisionalized structures and that this was independent of such factors as size. Research evidence is not, however, conclusive here and it has been shown that many firms have adopted M-form structures when not employing a diversification strategy (Rumelt 1982). Research has also shown that some firms which have followed a diversification strategy have not divisionalized (Rumelt, 1974).

It also appears to be the case that motivations for the move to M-form structures were different for companies in the USA than for those in Europe. For example Franco (1974) has argued that strategies of diversification may be a more important reason in the USA than in Europe. Franco argues that, in the European case, the emergence of product diversification may not in itself have been responsible for structural changes toward M-form companies, but that the changes in structure witnessed in Europe may have been a response to a less 'negotiable' environment. Since World War II, and especially since the creation of the European Community, the ability of firms to negotiate their environment has declined: tariff barriers have been reduced and competition, with both European and outside firms, has increased as has the enforcement of anti-trust regulations. Franco argues that, in industries whose environment has become less negotiable, the movement toward divisional structures has been the strongest (and, conversely, that where the environment is still negotiable, there is little structural change even in diversified firms). Product diversification may not therefore, be the only strategy that leads to multi-divisional organizational patterns.

Practising managers have also been quick to point to a string of operating advantages for the M-form structure. No doubt these 'reasons' (sometimes developed in propaganda-like fashion) have contributed to its diffusion. For example, many managers point to the greater 'bottom line' orientation of the M-form company and it is commonplace to assert that in functionally organized companies the focus tends to be on the work required to produce the results, rather on the results themselves. Thus the creation of M-form companies is associated with a move to emphasize the product market and the consumer, while larger functionally organized firms are more producer- or production-orientated. This consumer orientation lies behind attempts to introduce M-form structures and behaviour into the public service sector in the UK. It has also often been suggested that the divisional product structure promotes 'clarity of purpose' (as all employees within a particular division work toward a desired end result). Still other managers have suggested (somewhat ironically) that control possibilities are actually enhanced within divisional structures. Division management can be held accountable for the consequences of its actions because results and responsibility can be more easily identified. In the M-form structure, profits, revenue contribution, and costs can all be more easily determined by product or product group. Finally, many managers we have spoken to point to 'internal motivational' reasons for employing M-form structures. In the divisional structure each manager is responsible for his/her own patch and can see the results of his/her own actions. 'Feed-back', 'completeness', and 'autonomy' tend to define more fully the manager's job in an M-form structure than in a functional structure (Jackson 1986: 179). Each division manager must learn how to run a total business, not just one functional aspect of the organization (ibid.).

The strategy literature appears to offer a wide-ranging set of reasons, sometimes seemingly contradictory, for the development of the multi-divisional company. In their attempts to explain its emergence and growth, many researchers (Fligstein 1985; Clegg 1990; Mahoney 1992; Palmer and Zhou 1993) have organized these diverse explanations according to the underlying theoretical position of the analysis. The idea here is that different theoretical perspectives or conceptual lenses will lead to different explanations about the growth of the multi-divisional company (even when working from the same data set). The literature has generated (at least) five explanations relating to the growth of the multi-divisional company, each of which derives from a different theoretical position. Each of these is briefly summarized below. (For a more extensive summary of these positions the reader should refer to Clegg (1990) or Palmer and Zhou (1993). Our own categorizations are based around these. We do not discount or champion any particular explanation, but feel that each perspective or theory offers important motivating reasons for the growth of the multi-divisional company.)

1. *The Strategy–Structure Theory.* This explanation is organized around Chandler's (1962) well-known 'structure follows strategy' theorization. Chandler argues that a move towards product diversification strategies pushed or induced organizations to structure along divisional lines. A strategy of diversification is seen to lead to problems of accountability, control, and co-ordination which can only be resolved by the move to a multi-divisional structure. For example, an increased number and variety of product lines will overload the decision-making capabilities of centralized, functional organizations (Galbraith 1972). Reorganization from the unitary/functional organization to the multi-divisional company minimizes information overload problems (Mahoney 1992: 51).

2. *Transactions Cost Theory.* This theory is based around Williamson's (1970) work on increasing co-ordination costs in

growing organizations. Simply put, it is argued that the overhead costs of co-ordinating transactions in growing organizations lead to the development of the multi-divisional structure. As unitary or functional organizations grow in size, loss of control is caused by (1) fragmentary or erroneous information moving up the hierarchy, (2) inadequately operationalized instructions as they move down the hierarchy, or (3) the intentional falsification of information by lower level managers. From this perspective the crucial independent variables are measures of growth and size.

3. *The Population Ecology Theory.* Borrowing from an array of biological analogies and Darwinian arguments this theory suggests that transformations in organization form, such as the development of the M-form company, are more likely to occur through the creation of new organizations rather than through transformations in older, existing ones. As organizations age they become resistant to change and less likely to change their formal/structural properties. Thus, according to this theory, the divisional structure is more likely to be found among younger organizations. Hannan and Freeman (1977, 1984) have provided the groundbreaking analysis in this area.

4. *Political Theory.* According to this theory, explanations based on efficiency hold little water when we try to understand the movement to multi-divisional structures. Structure follows neither formal strategy nor particular efficiency imperatives (such as overhead co-ordination costs). Instead, structure is a product of both internal and external politics. Dominant coalitions maintain or introduce new structures that bolster their own power, even if it means that efficiency declines or profits drop. For example, Fligstein (1985: 380) has argued that the multi-divisional firm 'could be viewed as a mechanism which allows for growth through product-related and unrelated strategies, its implementation would be favoured by those who stood to gain the most from these strategies, i.e., sales and marketing, finance and personnel'. Thus, it is suggested that the multi-divisional firm will be introduced only if the structure favours the interests of the leading (or ascendant) coalition(s). Our own research activities often became embroiled in organizational politics, which leads us to suggest that all the explanations provided must have an element of uncertainty and politics must play a part as a motive force for change.

5. *Institutional Theory*. This theory suggests that organizations are induced to create a structure that looks like that of other organizations (in similar fields) in order to appear 'legitimate', 'modern' or 'up-to-date'. Institutional theory also eschews profit/efficiency reasons (except at a time when fields are founded or reorganized). What is important to remember here is that, according to institutional theory, organizations which have taken on the multi-divisional structure have done so principally because they want to look like other 'progressive' organizations (which might help to explain the current fashion in the public service sector). The work of Meyer and Rowan (1977) has laid the foundations for this theorizing.

CONCLUSION

The spread of the multi-divisional company has marked an important development in modern business history. It emerged early in this century as an alternative to the unitary or functionally organized company (which grouped tasks according to function and centralized most decisions). The multi-divisional structure is distinct from the unitary or functional company in its attempt to configure tasks according to the product or geographic markets to which outputs are directed and in its interest in separating tactical from strategic decision-making. The multi-divisional company allowed corporations both to diversify widely and to expand size and scope dramatically (Chandler 1962). Both the extent of the proliferation of the multi-divisional company and reasons for the growth of this particular structure were considered in this chapter. In the next chapter, we turn to a consideration of the strategy construct and ultimately to a discussion of the relevance of the construct for the multi-divisional company.

3
The Strategy Concept

This chapter provides a broad-based view of the strategy concept. The first part looks at strategy in a general sense, locating it within a traditional 'corporate' level of analysis. In the next section, the level of analysis is shifted from a corporate level to a functional level. Here, strategy-making is considered within the human resource management function itself. Thirdly, the key links between corporate and human resource management strategy are specified and outlined as developed or hypothesized in the recent literature. Fourthly, the notion of 'strategic choice' is highlighted along with the relevance of 'choice' for our own ensuing analysis of strategy-making in the multi-divisional company. Finally, we provide some preliminary perspectives on the link between corporate and human resource management strategy within the context of the multi-divisional company.

STRATEGY

The terms strategy and strategic management are widely used in descriptive and prescriptive management texts. Strategy presupposes importance and, in the words of the Concise Oxford Dictionary, 'generalship'. Thus strategy is associated with the long-term decisions taken at the top of the enterprise and is distinguished from operational activities. While it is generally true in multi-divisional companies that the corporate office is concerned with strategy and business units with operations, this can be misleading. An important part of the activity of head office is the control and monitoring of business unit performance, ensuring that budgets and performance standards are met. This can lead to a situation where corporate managers are actively involved in operational day-to-day activities. At the same time, divisions and business units may be given responsibility for the formulation of their own business strategies such as new product development, major investment proposals, and even take-overs, acquisitions, and disposals. A simple, hierarchical view of strategy is thus unhelpful. A more fruitful approach is to look at the attributes of management decisions in terms of importance or 'bigness' and the varying nature of the decision-making process in terms of its complexity or simplicity. In reviewing the literature on strategic decisions in management, Johnson (1987: 4) suggests that they are likely to be concerned with some of the following:

1. the long-term direction of the organization;
2. the scope of the organization's activities;
3. the matching of organizational activities to the environment;
4. the matching of the organization's activities to its resource capability.

Strategic decisions are thus 'likely to have major resource implications for an organization'.

The nature of decision making in strategy differs from that of the day-to-day activities of managing. Strategic decisions are likely to involve a higher degree of *uncertainty*. The reaction of competitors, governments, or trade unions to a given decision will be unknown in the medium term. Thus, uncertainty is associated with risk-taking. Strategic decisions are likely to be risky, and, given their long-term nature and the extent to which major resources are committed, it is likely that pay-off periods will exceed decision-makers' ability to foresee the future in anything but hazy outline. Secondly, strategic decisions are likely to demand an *integrated* approach to managing the organization. 'Thus a strategic decision and the implementation of strategy are likely to involve managers crossing boundaries within the firm in negotiating and coming to agreements with managers in different parts of the business.' (Johnson 1987: 5–6) This is of particular importance in employee relations, since initiatives taken by the corporate personnel staff have to be integrated through agreement and negotiation with often more powerful corporate colleagues and enacted at a much lower level by managers over whom the personnel specialist has little control.

The third distinctive feature of strategic decisions is that they are likely to be concerned with *change*. The opportunities for change, and the perception that change is required and is possible, are not continuous but periodic. Attention will need to be focused on the opportunities for change, the triggering of an awareness that change is required, the mobilization of resources and power behind the proponents of change, and on the process of searching for, selling, and implementing an alternative course of action.

Implementing change of strategic dimensions (i.e. decisions with long-term effects) is 'likely to involve the persuasion and organization of people to change from what they are doing. Moreover, the expectation may be that they change towards something that is ill-defined, uncertain and unfamiliar. Not surprisingly the management of strategic change can be highly problematic.' (Johnson 1987: 6)

The concern of this book is with strategic decisions which appear to affect the management of people in the enterprise. A priori we assume that long-term decisions within the ambit of employee relations will be of interest in so far as they set the fundamental patterns of labour–management interaction. Beyond this there is a need to examine the way in which multi-divisional companies are structured to achieve their long-term aims and how strategic decisions on internal operating procedures impact on employee relations.

STRATEGIC HUMAN RESOURCE MANAGEMENT

The 1980s were pivotal, for both academics and practitioners, in the coupling of the strategy concept to the human resource management function. Many of the standard texts within the human resource management field are now prefaced with the word 'strategic'. This coupling of the concept to the function is itself very much linked to the development of the personnel function and pressures to 'professionalize' and elevate the status of the personnel/human resource management department. In fact, it has become commonplace to make a distinction between 'personnel management' and 'human resource management'. For example, personnel management is commonly associated with traditional, non-strategic functions such as hiring, firing, and record maintenance, while human resource management is associated with the more sophisticated strategic management of employment relations. Each term is easy to parody. Personnel is a rag-bag function concerned with perceived 'low level' activities like personnel records, welfare, training, and salary administration; human resource management is becoming the fashionable title for the executive who wishes to portray him or herself as the organizational polymath contributing to strategy by added-value people-centred policies. Beneath the parodies are some fairly serious distinctions that relate to the focus of the subject and the level of the employee covered by the term.

Personnel management has a chequered history and has suffered from an accumulation of often routine administrative functions as organizations grew in size and specialization and the need for bureaucratic forms of control developed in response to legislation and labour market requirements. By and large the prime focus of personnel is the individual employee and labour-force resourcing for the organization. It has historically been much more concerned with the supply of labour than with demands placed on individuals in terms of their jobs and task. Its remit covers and matches the life-span of the employee from recruitment and selection, through training, health and safety, payment, and job transfers to termination and retirement. The design and administration of systems associated with these functions plays a large part in routine personnel activity. It is focused more on non-managerial staff and workers than on managers themselves, if only because these people normally constitute a majority of employees. Base disciplines are not particularly important but include industrial sociology with a marked bias towards the human relations school, social psychology, individual labour law and, increasingly, computer studies as information technology is applied to the area.

The term human resource management (HRM) originated in the USA in the sophisticated non-union companies which developed initially in the electronics industry, like Texas Instruments, Hewlett Packard, and IBM, and spread later to other non-union companies (Foulkes 1980) and to the new, non-union plants of traditional firms. The most commonly used definition of human resource management in the UK is that set out by Guest (1987). Human resource management, he says, 'comprises a set of policies designed to maximize organizational integration, employee commitment, flexibility and quality of work'. (Guest 1987: 503)

Guest has itemized four essential goals that are implicit within human resource management relating to integration, employee commitment, flexibility and adaptability, and quality.

1. On *Integration*. Human resource planning must become an integral component of the strategic planning process; must cohere with marketing, production, and finance, must be recognized by line managers as their responsibility as seen in their attitude and behaviour towards their employees, and employees themselves should be integrated as fully as possible into the business, so that 'what is good for company is perceived by employees as also being good for them'.

2. On *Commitment*. The requirement is for both a commitment to the organization and for job-related commitment. This requires employees to have attitudes which show 'strong acceptance of and belief in an organization's goals and values; willingness to exert effort on behalf of the organization and a strong desire to maintain membership of the organization'.

3. On *Flexibility and Adaptability*. An avoidance of rigid, hierarchical, bureaucratic structures, extensive decentralization and delegation of control, careful design of jobs, and flexible jobs undertaken by flexible employees.

4. On *Quality*. A requirement to recruit, develop, and retain staff with high levels of ability, skill, and adaptability; a requirement to set and maintain high standards and get them agreed or internalized by subordinates; the development of a public image of the organization as one with a reputation for high quality in its treatment of employees, such that the right sort of individuals are attracted to work for it.

This approach to human resource management with its emphasis on employee development and empowerment is sometimes described as 'soft' HRM (Storey 1992). The contrast is with 'hard' HRM where the concern is with resource utilization and cost minimization as discussed in Chapter 7.

These 'soft' goals are largely unattainable for most companies and portray an image of the world of work which is unrecognizable to many people in most sectors of employment. It may well relate to the way some firms manage their managerial and professional staff through management development, flexible career structures, retraining programmes, performance-related pay, share option schemes, appraisal systems, goal-setting, and very careful selection using psychometric screening and assessment centres. These techniques are growing in popularity in management development, but elsewhere there is little hard evidence of the growth of human resource management. Even green field sites have experienced difficulty in implementing and operating these principles, despite the best of intentions (Newell 1993).

A sustained debate is now in full swing on the meanings attached to human resource management in both theory and practice (see for example Storey 1989, 1992), and the connection with industrial relations (Boxall 1993; Purcell 1993*b*). It is not the purpose of this book to muddy the waters yet further if only because the practice of human resource management is more often found to be closely associated with manufacturing or operations strategies, not with corporate strategies. That is to say, it is linked to movements like Total Quality Management, which concern both the supply and demand side at the point of production. The activities of the corporate office may, and often do, limit or enable operational managers in the extent to which they can implement human resource management policies to suit their operational requirement. Thus, when it is argued that human resource management is closely linked to strategy, what is meant is business, operational or functional strategy, not directly *corporate* strategies, except in general terms of 'style' or 'culture' as discussed in Chapter 7. It is clear that not all firms will have, or choose to adopt, soft human resource management practices, whatever title they put on the door of the personnel director, yet all firms have employees. Our own view is that, strictly speaking, human resource management is best seen as a distinctive set of policies such as those described by Guest and by Storey, which might be appropriate for some firms in some circumstances.

LINKS BETWEEN CORPORATE AND HRM STRATEGY

The term human resource management has itself been increasingly coupled to the strategy concept (in fact, many would argue that you cannot have one without the other). In this sense, strategic human resource management is chiefly about integration of human resource management with corporate strategy and the strategic needs of the business. It is for this reason that some (Schuler 1992: 30) argue that successful efforts in strategic human resource management begin with the identification of strategic business needs. This integration theme is central to virtually all of the current definitions of strategic management. Thus, strategic human resource management is usually defined as the integration of the policies and practices of managing employees with the corporate strategic plan of the organization. To provide some concrete examples here, it has been suggested that both staffing practices (Schneider 1983) and career development practices (Olian and Rynes 1984) should be linked to the long-term goals and strategies of the organization. The value of linking an organization's human resource management strategy to its corporate strategy has been pointed out by many scholars (Tichy et al. 1982; Tichy 1983; Dyer 1984; Alpander and Botter 1981). The literature also tends to be confident that close links will lead to greater competitiveness and improved organizational effectiveness.

Much of the research work in the human resource management strategy area has revolved around the nature of the links between corporate and human resource management strategy. A variety of links have been identified and it is possible to classify these in terms of a 'reactive–proactive' continuum (Kydd and Oppenheim 1990: 147). In the 'reactive' mode, human resource management strategy is seen as fully subservient to corporate strategy, and corporate strategy ultimately determines human resource policies and practices (see Walker 1981, 1988; Schuler 1992; Schuler and Jackson 1987 as examples of such theorizing). Once the corporate strategy is set (independent of human resources input) human resource policies are implemented to aid in the delivery of the strategy. An example of this type of reactive model can be found in Schuler's work (1992: 20) in Figure 3.1. The positioning of organization strategy at the top of the figure and the one-way arrows downwards are no accident: corporate strategy and the strategic needs of the business are seen as the prime determinants of human resource activities. The practice of strategic human resource management in this sense is all about aligning the 'five Ps' (philosophy, policies, programmes, practices, and processes of human resource management) in a way which will enhance strategic initiatives. The management of people thus

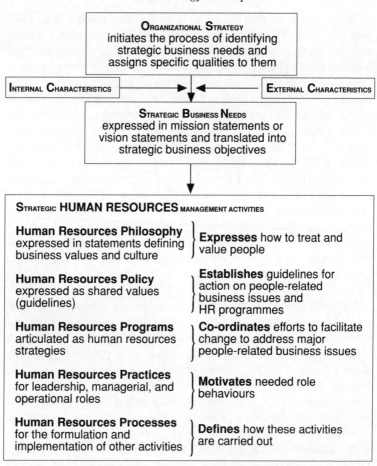

FIG. 3.1. Schuler's Model of Human Resource Strategy
Source: Schuler (1992: 20).

becomes a residual activity. The reactive linkage can be depicted in a schematically simplified way as follows:

corporate strategy ------▶ human resource strategy

In the 'proactive' mode, the human resources function is itself actively engaged in the strategy formulation process (see Dyer 1988, 1984 as an example of this school of thought). The extreme position in this mode is seeing human resources as the determinant of corporate strategy itself (the strategy in this sense is the people: who they are, their vision, their collective will, etc.). Peters and Waterman's *In Search of Excellence* (1982) made an early call for a more interactive linking between corporate and human resources strategy. We can depict this interactive linking as follows:

corporate strategy ------▶ human resource strategy
human resource strategy ------▶ corporate strategy

In addition to the corporate strategy–human resource management links identified above, the literature has also pointed to the importance of the environment as a mediating variable. Bamberger and Phillips (1991) depict this three-way link, as shown in Figure 3.2. Those models which do incorporate the idea that the environment serves as an important determinant of human resources activities (Wills 1984) tend, however, to offer little insight in terms of the precise nature of the interaction between corporate strategy and the environment in the shaping of human resources strategy (Bamberger and Phillips 1991: 159). At best, the nature of the interaction is seen to be limited to the role of business strategy in mediating the impact of the environment on human resources strategy (Tichy *et al.* 1992).

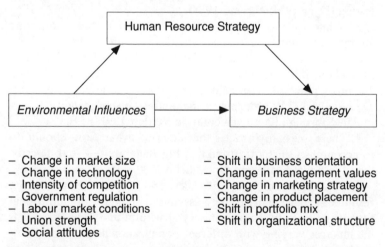

- Change in market size
- Change in technology
- Intensity of competition
- Government regulation
- Labour market conditions
- Union strength
- Social attitudes

- Shift in business orientation
- Change in management values
- Change in marketing strategy
- Change in product placement
- Shift in portfolio mix
- Shift in organizational structure

FIG. 3.2. Environment as a Mediating Variable for Human Resource Strategies
Source: Bamberger and Phillips (1991).

We should note, moreover, that this lack of precision and detail in the nature of the environmental links also holds for much of the writing on the strategy–human resource management linkages we have discussed above. Indeed much of the work on the nature of linkages has been developed at an abstract and highly generalized level. While the current strategy–human resource management literature has been of value in pointing to the need for businesses to pay greater attention to the strategy–human resources linkage and in mapping out generalized linkages, it has left many questions unanswered. It is possible to point to a number of limitations and problems with the current literature and we summarize these now (for the limitations and problems really serve as the motivation for our own work):

1. The specified links between strategic planning and human resource management have been too theoretical and abstract. The generalized models, while compelling in their logic, provide little for the practitioner at a more detailed operational level. At best, existing models serve as useful evangelical calls for attention to this area.

2. The few prescriptions which have been identified have been found to have limited validity when subjected to careful empirical testing (Dyer 1984; Nkomo 1988).

3. While various links have been specified between corporate strategy and human resources activities very little is known about the precise nature and content of these interactions. We know little about the specific steps required to link strategy and human resources practices, especially at the corporate level where strategy often lacks clear guiding themes and is 'to some extent . . . formulated through its implementation' (Grant 1991: 359).

4. There are many articles that describe 'what' firms should do to accomplish this critical linking, but there are few that describe 'how' to do it (Kydd and Oppenheim 1990: 145).

5. The research into links between corporate and human resource management strategy assumes a standard organizational form (usually a functional/unitary structure) and does not imagine linkages varying with different organizational structures. How might links differ in a multi-divisional company from, say, those in a functionally organized company?

6. There has been an inadequate theorization of the more basic

aspects of corporate strategy (e.g. the nature of product markets) and their links to human resources strategy. We need to know more about basic issues like the link between markets and the management of employee relations (Marchington, 1990: 111). The furthest that the literature goes is simply to assume that the product market determines managerial style, or at least severely limits the range of options open to management.

7. We have a suspicion that the literature has been too optimistic about the possibilities for the 'close' (and harmonious) linking of corporate and human resources strategy. Schuler (1992: 20–1) himself points out that the links between human resources activities and business needs tend to be the exception even during non-turbulent times. When such links do occur, they are usually driven by the organization's efforts to formulate and implement a particular strategy.

8. The current human resource management strategy literature tends to be based on 'good management' practice and a rational model of both 'man' and organization. Such a model discounts the significance and power of both politics and culture as variables in shaping (and possibly subverting) strategy.

THE STRATEGIC CHOICE PERSPECTIVE

Our own book represents an attempt to respond to the above set of shortcomings of the current literature. In an attempt to make the strategy area more concrete, we prefer to look at it in terms of 'strategic choice.' The base assumption of the book is that the multi-divisional (M-form) company has a unique set of strategic choices open to it in the management of human resources. The use of the term strategic choice puts emphasis on the 'flexibility that management has to identify future interests and to exercise choice' (Timperley 1980). The literature on strategy and strategic choice in the M-form company has tended to neglect those choices relating to the management of human resources and personnel. The growing literature on strategy in industrial and employee relations (Timperley 1980; Thurley and Wood 1983; Kochan, Katz, and McKersie 1986), on the other hand, has rarely differentiated between organizational forms and we therefore know little about employee relations strategies particular to organizational structures such as the M-form company.

Strategic choice contrasts with the deterministic model of organization in which actors are so severely constrained (by environmental factors such as organizational size, technology, product markets, etc.) that there is little or no choice open to management. When applied to employee relations, the notion of strategic choice highlights the distinction between traditional personnel management, concerned with 'fire-fighting' and mere reaction to union and employee pressures, and personnel management concerned with influencing employee relations outcomes through strategic decision-making. The employee relations system, within a strategic choice model, is not accepted as a given, but rather can be actively adjusted by managements which take key decisions that either directly or indirectly have human resources consequences.

The strategic choice perspective has been well put by Marchington *et al.* (1992: 44) who see there to be a continuum between determinism (where there is no choice, or no real choice) and autonomy (where the firm is free to choose what it wishes). Strategic choices must be set within a context of constraints such that

> the relationship between external and internal factors is seen as . . . complex. On the one hand managerial choice remains constrained by the environment within which the company operates, especially in the short-run and at lower levels in the managerial hierarchy. On the other hand, however, senior managers are treated as having considerable room to manœuvre, and their decisions also to some extent shape the environment within which the organization operates; for example, its markets, products, location, technology, size of establishment, and type of employee to be recruited.

The strategic choice perspective is more than just a reflection of the fashion in ideas. The turbulence of environment forces companies to make choices in areas where they had previously made standard assumptions about the way to do things.

A typical list of factors associated with environmental turbulence would include:

- the rapid diffusion of new technology;
- globalization and the creation of the Single European Market;
- mergers and acquisitions;
- privatization;
- the drive for competitiveness in the public services sector;

- enhanced knowledge of competitor practices through benchmarking;
- the progressive withdrawal of institutional regulation in the labour market;
- demographic changes;
- increased uncertainty over macro-economic conditions;
- the growth of the international capital market;
- the rise of institutional shareholders;
- an explosion in means and methods of information technology and communication systems.

This list is far from comprehensive, but the message is clear. It is very much harder to choose the well-known option of 'do nothing' which means, of course, do the same old thing without thinking about it. Choice is forced as much on reluctant managers as it is on organizational innovators. The former rely on the past as the guide to the future; the latter exercise strategic choice.

The advantage of emphasizing the notion of choice in understanding the concept of strategy as it applies to employee relations is that it brings substance to the debate where it has not existed in the past. Many past definitions of strategy (Hendry and Pettigrew 1986) have been kept at an exceedingly abstract level. In Hendry and Pettigrew's case, the high-gloss terms of 'planning' and 'coherence' somewhat disguise what they later identify as the component parts of strategic human resources management. When forced to relate general concepts to specific processes, we find that Hendry and Pettigrew's strategic human resource management equates for the most part to manpower planning and training and development, activities that have been around as long as personnel management itself. Rather than defining strategy in employee relations in terms of discrete activities (like manpower planning) we prefer to define it in terms of choices; fundamental choices which, when made, significantly alter employee relations outcomes. Strategic activity in employee relations begins when managers step back from taken-for-granted structures and systems and think about the possibility for change. This is contrasted with non-strategic thinkers: the executive who simply maintains given structures or makes assumptions about the nature of employment and the organization of work.

Various attempts have recently been made to describe and catalogue the key or important decision areas within industrial and employee relations (Kochan, Katz, and McKersie 1986; Dyer 1988). Dyer describes the major dimensions of organizational human resources strategy in terms of goals (labour costs, productivity, employee competence, etc.) and means (organization or work structure, employee utilization and allocation, etc.), while Kochan, Katz, and McKersie (1986) produce an industrial relations strategy 'matrix' which relates the level at which decisions are made (macro- or global level, employment relationship, and workplace) to the actors making the decisions (they see that, in addition to management, both unions and government can and do make strategic decisions, but maintain that the most important of these strategic choices have been those made by *management*).

Descriptions and catalogues of the key decision areas are important not only for prescriptive purposes (by at least pointing to those areas which can act as pivotal points for change) but also as building blocks for the development of a more general model of change in employee relations. There are problems associated, however, with producing descriptions or lists of key decisions, not least of which is that it would be extremely time-consuming to enumerate all the possible strategic choices open and available to management in employee relations. Lists like Dyer's, which fail to discriminate between various choices in terms of significance, reach the point of blandness in which strategic choice is reduced to a kind of a state-of-the-art personnel management textbook.

Indeed it is not surprising to note that some of the lists of choices in employee relations have in fact pre-dated the 'strategic choice' literature and can be found within the personnel management literature; one example is Thomason's list (1976), which includes: (1) the general question of the structure of negotiation to be encouraged in a firm; (2) the issue of the processes or procedures to be developed to facilitate order within the employment relationship; and (3) the particular issue of how the firm might react to various statutory obligations.

In enumerating a list or catalogue of choices in employee relations management it is important to isolate those choices that are significant and fundamental, that is, those that have the possibility of significantly altering current employee relations outcomes. Such decisions are 'big' decisions and in a way correspond to Kochan, Katz, and McKersie's macro- or global decision level, in which management intervenes in the employee relations and industrial relations system through its strategic role in plant investments, plant location decisions, union recognition, new technology, controls over outsourcing or subcontracting, and so on. The difficulty associated with the development of a catalogue of key choices lies in defining the criteria for what constitutes and what does not constitute a key choice. The decision as to whether a firm will manage its employee relations with or without trade unions is clearly a significant decision (that will set the tone for the management of all parts of its employee relations); a decision to alter working hours clearly has less significance and could hardly be classified as a strategic choice.

Some of these big decisions fall within the bailiwick of personnel management and have a direct personnel basis (such as union recognition), while others will be the prerogative of the board of directors or senior line management and will be made only partly for personnel reasons (such as a plant location decision). As Timperley (1980) notes, 'many of the major decisions to be made in industrial relations . . . are themselves contingent on other higher levels and contexts which do not necessarily appear as industrial relations decisions at all'. Regardless of who owns them, however, each of these decisions shares the common characteristic of 'bigness' and has the potential of significantly changing the make-up of the employee relations system. Strategic choice for industrial relations involves, therefore, choice by both personnel and line management.

We have suggested that it will be important to identify the more significant and fundamental choice areas. It will also be important to identify those choice areas which are most significant for particular organizational forms. Clearly as organizational form varies so will the range of strategic choices. Strategic management of employee relations may have, in fact, special significance for the M-form firm. As Thurley and Wood (1983) note, 'the argument for an industrial relations strategy in a firm may be based on the need to accommodate different managerial objectives and definitions of the situation and hence to reconcile conflicting managerial behaviour, especially in multi-plant situations'. It would appear to be the case, further, that complex organizations, at least in the multi-plant form, pay greater attention to industrial relations than do single-plant companies. Purcell and Sisson (1983) have suggested that: 'it seems not unreasonable to suggest that the structural changes identified above [to more complex organizational structures such as the M-form company] have led to an increase in the specialisation of industrial relations management, the development of policies and initiatives and a more lively debate within the ranks of senior management about the "union question" which previously had often been delegated to management at plant level.'

The M-form configuration not only appears to require some element of strategic choice but it also provides a rather unique and complex stage upon which strategy takes its course. Strategy in the M-form company occurs and is experienced at the three levels of corporate headquarters, the division and the operating subsidiary or unit. The communication, co-ordination, and control of a strategic plan become necessarily more problematic. In M-form companies there is, in theory at least, a separation of profit responsibility at division level or business unit from strategic management at corporate level (Chandler 1976). Due to the political nature of organizational life this separation may not be this clear-cut, as strategy formulation (e.g. competitive strategy) may be just as much a part of divisional life (Porter 1987), and executives in lower levels of the organization can themselves exert upward influence on strategy formulation.

THREE LEVELS OF STRATEGY

The distinctive feature of multi-divisionals is that their internal operating procedures are more refined and differentiated than those found in functional or holding companies. The decision to move to a multi-divisional structure from, say, a centralized functional firm, or to adapt the configuration to emphasize local profit centres, is a strategic decision of substantial importance in its consequences for employee relations, as discussed in Chapter 4. The decision to reorganize might have been triggered by strategic decisions taken earlier, for example, to diversify. One useful way of distinguishing between types of strategic decisions is in terms of upstream and downstream. Upstream, first-order decisions are concerned with the long-term direction of the enterprise or the scope of its activities. Clearly these decisions will have implications for the type of people employed, the size of the firm, and the technology required. If an upstream decision is made to acquire a going concern, a second set of considerations applies concerning the extent to which the new firm is to be kept apart from or integrated with existing operations, and about the nature of the acquired firm's relationship with its new owner. These can be classified as more downstream, or second-order strategic decisions. This is similar to Chandler's distinction (1962) between strategy and structure and his oft-quoted dictum that structure follows (i.e. is downstream from) strategy. The difference here is that decisions on strategy (the type of business undertaken now and in the future) and on structures (how the firm is organized to meet its goals) are both of strategic importance in that they have long-term implications for organizational behaviour, are taken in conditions of uncertainty, and commit resources of people, time, and money to their attainment; that is, a political model of decision-taking is required.

It is in the context of downstream strategic decisions on organizational structure that choices on employee relations structures and approaches come to be made. These are themselves strategic since they establish the basic parameters of employee relations management in the firm, but are likely to be deeply influenced by first- and second-order decisions as well as by environmental factors of law, trade unions, and external labour markets. These are termed here, therefore, third-order strategic decisions and apply to other functional strategies in, say, marketing, production, buying, sales, and so on. At its simplest therefore three levels of strategy are evident as seen in Figure 3.3. In theory, in this idealized model, strategy in employee relations is determined in the context of first-order, long-term decisions on the direction and scope of the firm's activities and purpose (location, technology, skill requirements, etc.), and second-order decisions on the structure of the firm seen in the context of its internal operating procedures (levels of authority, control systems, profit centres, etc.). What actually happens in employee relations will be determined by decisions at all three levels and by the willingness and ability of local management to do what is intended in the context of specific environmental conditions.

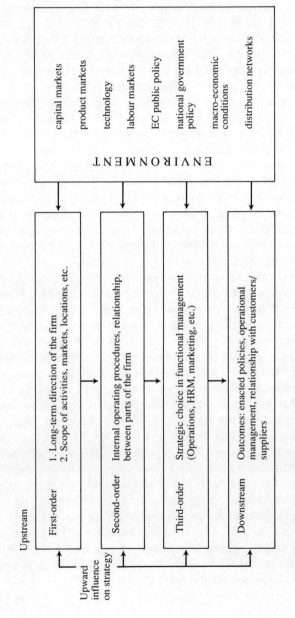

Fig. 3.3. Three levels of strategic decision-making

The Strategy Concept

One principal objection needs to be raised at this point on the nature of the model (there will be more raised in subsequent chapters). Like many such models it implies rationality in the process of decision-making: a carefully planned series of decisions where employee relations are designed to mesh with organizational structures which in turn derive from first-order strategy. If strategic decisions are characterized by the need to cope with change and uncertainty and to integrate management activity in various fields, then a political process model is more appropriate. 'Strategic decisions', writes Johnson, 'are characterized by the political hurly-burly of organizational life with a high incidence of bargaining, a trading off of costs and benefits of one interest group against another, all within a notable lack of clarity in terms of environmental influences and objectives' (1987: 21).

We must also allow for the fact that all strategic decisions are taken in historical circumstances in the sense that there is a prevailing pattern of action within the firm. This pattern of behaviour and expectations, especially in the area of employee relations, will itself exert influence over higher-level strategic decision-makers such that in some cases structure will determine strategy (as Hall and Saias (1980) put it). A given organizational recipe or paradigm may exert extraordinary influence over the minds of the top executives. Within the field of industrial relations this process can become yet more difficult, since trade unions provide a special type of extra interest group which does not necessarily subscribe to the goals of the firm. The process is further complicated by marked ambiguity in the area of personnel management. Since it is difficult to determine the ends (what is the purpose of personnel management?) and the means to achieve these uncertain ends, it is also difficult to measure whether the firm is successful in its personnel or human resources policies.

In a political model of strategic decision-making questions of legitimacy and power become critical. We will thus need to consider where power lies, how it comes to be there, and how the outcome of competing power plays and coalitions within senior management are linked to employee relations. As a foretaste, it can be argued that 'personnel', either the function or the department, is especially ambiguous because so few companies have deep-seated positive philosophies or guiding principles of any meaning which provide solid foundations on which to base third-order strategies. It may be possible for personnel executives to design a sophisticated employee relations strategy in isolation, but this may be at odds with first- and second-order business policy. It looks good on paper, is widely reported and discussed within the personnel profession but remains unrealized and unenacted unless the originators of the strategy gain the support of other senior management. Pettigrew (1975) has argued that power is a resource. Power resources might include control over information and systems, access to higher authority, group support, or specialist expertise. Personnel managers rarely have power over unique sources of information and systems and their specialist expertise is rarely recognized by other managers, most of whom will have pet notions of the best way to manage people. Employee relations strategies are thus unusually reliant on group support or access to higher authority for success. This support in turn will depend on the extent to which the strategies are seen to serve the needs of the business.

How do we relate and apply employment relations principles to the multi-divisional company itself? In the multi-divisional structure, it is likely that first-order business strategy will take primacy in setting the tone of human resource management. This is the case, because in both the conglomerate organizations (Tyson and Fell 1986) and the diversified firm (Sisson and Scullion 1985) the corporate personnel department tends to be poorly developed, leaving little room for central co-ordination in human resources strategy-making. The burden is placed on employee relations problems and issues. Without the benefit of a critical mass of human resource professionals, it might be supposed, and an attendant body of human resources knowledge and theory, there is less likelihood of the existence of a sophisticated human resource management culture and style. This may, however (as we shall see), not be a bad thing. In fact, it is possible to argue that human resource management in the multi-divisional company conforms more closely than anything else to Tyson and Fell's most advanced model of personnel management —the 'architect' model (1986). It is in this model that personnel executives seek to create and build the organization as a whole. This creative vision of personnel means contributing to the success of the business through explicit policies which seek to give effect to the corporate plan, with an integrated system of controls between personnel and line managers. Here the personnel function is most closely linked to business strategy and policy.

For the Guest model of human resource management to become meaningful within multi-divisional companies, questions as to the most effective ways to manage people at work have to be elevated in importance and form part of the first-order strategies on the long-term aims and purpose of the firm. Second-order strategies on internal operating requirements would have to take into account their effect on the employees' willingness to identify positively with the firm. The performance control systems designed to monitor and motivate unit managers' performance would have to include criteria designed to encourage a positive, supportive attitude towards employees. Finally, the enterprise would have to have a policy that it wished to recruit and retain the best quality of employee and be prepared to pay over the odds to attract and retain employees.

Some companies, such as Marks and Spencer, IBM, and Hewlett Packard, are well known for doing this and have long-standing corporate philosophies respecting the rights of individuals, although the recession of the early 1990s led each of them to abandon aspects of their policies. Most do not, and probably could not adopt all aspects of 'soft' human resources management; and, as will be discussed in a later chapter, it would require a considerable transformation in the importance attached to personnel questions, a reordering of business priorities and a shift toward a more centralized, unified enterprise for the full 'soft' version of human resource management to become widespread.

Before further progress can be made on the types of strategic choices that can be taken in employee relations we need to return to the question of the overall strategies, structures and styles found in multi-divisional companies. Choices about the management of labour can only be understood in the wider business context. How do corporate and business strategies affect employee relations? What is the connection between the internal operating procedures of multi-divisional companies and the structures of employee relations? How does the style of the corporate headquarters in managing subsidiaries and business units impact on the management of labour? These questions form the agenda for the next chapter.

SUMMARY

This chapter has been concerned with the strategy construct, both corporate and human resources strategy. Corporate strategy was defined loosely as those decisions which are 'big' in the sense of having a long-term effect on the behaviour of the firm or in committing major resources to a certain course of action. Strategic decisions were seen to be concerned with change, taken typically in circumstances of uncertainty, and to involve the integration and involvement of a variety of senior managers, with a mix of responsibilities and functional specialisms. As a result, we argued, the political nature of strategic decisions is emphasized, leading to questions of who has power in the organization. It was noted in passing that the lack of a specific system or information base in personnel management, coupled with the difficulty of generating specialist, unique expertise in 'people management', makes strategic decision-making in employee relations unusually dependent on gaining the support of other managers and/or access to and sponsorship of higher authority.

Three levels of strategic decision-making were identified. First-order, upstream decisions concerned the long-term direction of the firm and the scope of its activities; second-order, more downstream decisions concerned internal operating procedures and the relationship between the parts of the business. Strategic choices in human resource management were seen to be yet further downstream as third-order decisions. Thus an understanding of human resource strategy can only come about by looking at the opportunities and constraints imposed by first- and second-order strategies. The type of employee relations manifest in the firm at the shop-floor and in the office will be influenced by first-, but especially by second-order strategies, as well as by the strategic choices taken within the human resource area.

The next chapter is critical to the whole approach of this book since it sets out the nature of the multi-divisional company and how different strategies, structures, and styles influence employee relations and human resource management. For the moment we can summarize the discussion on the meaning of the term strategy as employed here by noting that:

- Strategy implies the making of decisions which have long term or 'big' consequences for the organization.
- Strategic decisions are typified by their uncertain outcomes, by the need to involve and integrate the views of a cross-section of executives, and by the fact that their purpose is to bring about change.
- Strategic decision-making is best seen as a political process, not the outcome of 'rational', value-free analysis. As such, questions of power and who holds power become critical.
- Three levels of strategy need to be distinguished along a continuum of upstream and downstream decision-making. Questions of long-term direction and the scope of activities set the context for decisions on internal operating procedures, notably the relationship between parts of the business. Specific employee relations and human resource strategies are third-order activities, deeply influenced by first- and second-order choices.
- What distinguishes diversified, multi-divisional companies from simpler, functional firms is the range of strategic choices open to them. They are not prisoners of their environment but are more able to shape and structure their activities and environment to maximize the achievement of their purpose, always supposing they have sought to define this in the first place beyond a vague desire to ensure survival.

4
Corporate Strategy and the Influence of Personnel

The object of this chapter is to trace the relationship between diversification business strategies, found typically within the multi-divisional firm, and human resource strategies and practices. The chapter begins with a general discussion of corporate strategy as it relates to the multi-divisional firm and then proceeds to work out the links between corporate and employee relations strategies within the context of the diversified firm.

TYPES OF STRATEGIES

One of the critical features of multi-divisional companies is the extent to which they have diversified the scope of their activities as part of their corporate strategy. A related distinctive feature of multi-divisional enterprises is that internal operating procedures are more refined and differentiated than those found in functional or holding companies. In the classic literature two types of distinction are important here. First there is a separation of corporate strategic planning and financial control from operational management at unit and/or divisional level (Williamson 1970) and, secondly, decisions on the structure of the enterprise, notably the relationship between the parts, follow from strategies to do with the long-term direction of the firm and the scope of its activities (Chandler 1962).

It has been suggested that strategic decisions are limited to those decisions taken periodically about the mix of businesses and long-term direction of the firm. This ignores the reality, however, that such decisions invariably cascade into a second-order stream of fundamental business decisions, arguably as critical as a first-order type of decision such as the 'mix of business'. The decision to acquire a business, for example, will lead to a second set of decisions about structures. To what extent, if at all, are synergistic economies of scope to be realized by integrating parts of the new business into the parent company, or is it to stand alone, unconnected, except in a financial and ownership sense, from the parent enterprise? The strategy of the enterprise does not, therefore, lead inevitably to a given structure, and indeed many firms have adopted structure strategies based on the decentralization of profit centres without altering their long-term aims or business mix, as for example in parts of the public services sector.

It is for this reason that we suggest that a distinction be made between what we refer to as *first-order* corporate strategy, concerned with the enterprise's basic goals and range of businesses and markets served, and *second-order* business strategies, focusing on the internal operating procedures which establish the nature of the relationship, and looseness or tightness between the corporate office, divisions, and operating subsidiaries. It should not be surprising that Hill and Pickering (1986) and Hill and Hoskisson (1987) have shown there to be a wide variety of structures within the generic form of the multi-divisional company. These are, in part, related to the type of economies the firm is seeking to maximize and the extent to which corporate management consciously seeks to structure the firm to maximize these economies. It is suggested that it is in light of these two levels of strategy that decisions on employee relations come to be taken in most multi-divisional firms.

Our notion of first-order corporate strategy and second-order business strategies could be said to fit with those who place stress on the 'significance' aspect of strategic decisions, that is those who see them as decisions which shape 'what happens for a long while after' (Hickson *et al.* 1986: 27). Strategic decisions in this line of thought are usually taken in conditions of uncertainty (in that the effects of the decision are unlikely to be known for a reasonable period), are likely to have major resource implications, and tend to involve more than one functional area of management (Johnson 1987). This 'significance' definition suggests that strategic decisions are something more than decisions about the mix of the business.

Decisions about the structure of the internal operating

procedures—the number of divisions, strategic business units, the location of profit centres, the type of performance criteria used—are clearly of strategic importance in that they provide the critical structures for organizational behaviour and performance. At the same time these decisions are of a different order of importance to larger corporate strategies to diversify, acquire, or dispose of parts of the business and set the long-term direction of the firm. Nevertheless they are also strategic in the sense that there is no one obvious choice to be made. The outcome of a given structural form is impossible to predict. Not surprisingly fashion and the pursuit of organizational politics both play a role in the determination of second-order decisions. Some companies seem to reorganize every few years, that is, they keep changing their second-order strategies.

Similarly, certain policy decisions in employee relations can themselves be strategic in that they commit the organization to a course of action for the foreseeable future, involve different levels and different areas of management, and often have resource implications. One could cite here decisions on union recognition, types of payment and job grading systems, labour force flexibility policies, and levels of collective bargaining. Decisions to change the location of collective bargaining within the enterprise or to open non-union, green field sites are, in these terms, clearly strategic. We refer to key employee relations decisions as *third-order* strategic decisions.

We argue below that it is possible to put forward an idealized or normative model of strategy in which decisions on the scope of activities and goals (what we term first-order strategy) will lead to decisions on organizational structure and internal control mechanisms (what we term second-order strategy) which, in turn, will create a dynamic pressure to fit employee relations strategies (what we term third-order strategies) to the needs of the business. This model leads to an image of cascading sets of strategic decisions in which upstream, first-order corporate strategies flow downward into second-order structural decisions and, further downstream, third-order employee relations strategies.

The first two links have been explored by Hill and Hoskisson (1987), who distinguish between financial, synergistic, and vertical economies in the way multi-divisional firms are structured. Financial economies (financial control) are found where the superior allocative properties of the internal capital market are maximized. That is to say that financial criteria dominate the goals set by the corporate office, compared with long-term growth or market development for example. These enterprises are likely to have diversified into unrelated activities and seek to structure the firm in such a way as to minimize or ignore interdependencies between business units and divisions. Synergistic economies exist where common techniques, skills, or market knowledge are utilized across a range of products. These companies are capable of organizing around key values and are usually concerned that the acquisitions match and enhance their business mission. Second-order integration strategies are likely here. Vertical integration happens where a closely co-ordinated and integrated chain of activities from raw material sourcing into final distribution allow vertical economies to be realized. Here there is likely to be a strong central planning influence and an emphasis on administrative co-ordination.

Hill and Hoskisson hypothesize that 'as firms grow by vertical integration or related diversification, they will become increasingly constrained by information processing requirements to focus on attaining financial economies, [and that] under conditions of either high or increasing uncertainty, vertically integrated firms will focus on realising financial economies' (1987: 338, 340). Thus the changing scope of the enterprise's activities as it grows, coupled with uncertainty in product and capital markets, tends to lead to changes in structure. Second-order strategy decisions are taken to emphasize the financial control and reduce the planning and administrative role of the corporate headquarters. Firms emphasizing the achievement of financial economies and financial control are 'characterised by relatively high degrees of decentralisation of decisions, decomposition between divisions (i.e. minimising relatedness and coordination) and consequently, high accountability for divisional profits' (ibid. 334).

As firms move to emphasize financial control they are likely to adopt elements of portfolio planning systems (Hamermesh 1986) which also focus on decentralization, unit autonomy (except in the financial sense), and the minimization of interrelatedness. Here the corporation's corporate strategy manages the mix of businesses held in ownership (the portfolio) in order to gain a balance across the business cycle (i.e. cyclical and counter-cyclical businesses) and in various degrees of maturity, so that declining business comes to be replaced by the new firms in the portfolio. Needless to say, each business is likely to be kept apart from the others. Accounting practices are likely to emphasize the separation and independence of each business, so that the contribution of each profit centre to the corporate balance sheet can be measured.

Not only are first- and second-order strategies linked, but we are also suggesting that third-order strategies will themselves be linked to the first two strategy levels. The question for this chapter is: precisely how are these linked, and in which direction? Much of our argument thus far has presumed a rational ordering of decisions from strategy to structure. In practice, the incremental, emerging, and political nature of decision-making can make the process seem far less ordered, such that second- (and third-) order strategies can be formulated in isolation from or in advance of first-order strategies. Different groups of managers are often involved in formulating these three levels of strategy and sometimes pursue quite different interests. This is especially the case in the frequent disagreements between headquarters staff and managers in the field, as seen over the centuries in the military.

Normative or idealized models of how the world ought to be also assume that the full consideration of the impact of strategy on all factors, functions, and areas of the business is made at the time of the strategic decision-making including, for our purposes, the human side of business. Thus, in theory, first- and second-order policy-making ought to include full consideration of their effect on employee relations, and be modified to allow for the maximization of the contribution of human resource management to the corporate goal. This would be the case both in the reactive model of human resource management (if only to ensure complete implementation is likely) and in the proactive model, where strategies are decided on in the light of human resource factors. As will emerge in the next section, we have cause to doubt that either type occurs in most firms.

It is also clear that, while shifts in first- and second-order strategies will create pressure for change in employee relations structures and procedures, such pressures will not always result in the appropriate adjustments because of vested interests within the employee relations sphere itself. We can question whether or not those firms moving towards a diversification, decentralization, and financial control strategy will a priori, be liable to exert strong pressure on employee relations strategies, the role of corporate personnel departments, and on the structures of the institutions and procedures of industrial relations, especially in those firms where single employer bargaining predominates (Sisson 1987). The actual effect of corporate strategies on the management of employee relations is discussed in the second half of the chapter.

THE INFLUENCE OF HUMAN RESOURCE ISSUES ON CORPORATE STRATEGY

What role does human resource management play in the strategic planning process? There appears to be a wide discrepancy between those (Miller 1989) who suggest a need for a central role, and actual research evidence (Winkler 1974, Hickson and Mallory 1981) which indicates a relatively weak position of personnel management in the corporate office, and outlines the limited extent to which personnel and industrial relations issues are considered in strategic decision-making. While Miller (pp. 348–9) for instance, would assert that 'the strategic management of human resources is a vitally important component in strategic management generally' there is, in fact, scant evidence to show that human resource issues are considered in strategic management and plenty of data to show personnel is a stand-alone function.

It might be expected that decades of public exhortation and the powerful, if dubious, anecdotal evidence of Peters and Waterman in *In Search of Excellence* (1982) would have ensured that human resource issues were taken into account in the oft quoted examples of 'people and productivity'. More recent evidence, which we report below, points to a general state of affairs where the influence of personnel management over strategic decisions is limited and where the management of employee relations is deemed to be an operational matter wherever it is located. The findings reveal little international variation. Hegarty and Hoffman, for example, studied 109 firms in western Europe drawing information from '407 top managers . . . [all of whom] were part of the top management team' (1987: 77). The purpose was to determine who influences strategic decisions. Aggregating the data over all decisions, personnel influence was rated well below other areas at the bottom of the scale with 'almost no influence'. The authors conclude that 'personnel's area of responsibility appears to be too narrow to exert influence over strategic decisions'.

In the USA Nkomo studied the Forture 500 firms and found that only 15 per cent used comprehensive systems of human resource management integrated with strategic business planning. 'Strategic human resource planning efforts appear to be carried out in isolation from strategic business planning, or human resources planning decisions are treated solely as a derivative of strategic business planning rather than a primary function' (Nkomo 1988: 66). Also in the USA, a study of diversified firms by Christiansen found that 'labour relations professionals are isolated in the sense that they tend not to be involved in critical decision making that does not explicitly involve union contract negotiation or administration' (Christiansen 1987: 375).

In Britain, studies of strategic decision-making confirm this general picture. Hunt and Turner (1987) showed that 'the human factor' tended to be ignored in mergers and take-overs; personnel issues came to the fore only after the decision was taken. Butler and his colleagues looked at strategic investment decisions and confirmed earlier work by the same team (Hickson *et al.* 1986) that, in taking investment decisions, the least attention was given to personnel issues (1987: 8).

Perhaps the best and most comprehensive piece of evidence, however, comes from the first company-level industrial relations survey (Marginson *et al.* 1988). First, there is the question of personnel directorships. It might be argued that membership of the board of directors would indicate a strong potential for due consideration to be given at that level to employee relations issues in strategic decision-taking. Overall, 31 per cent of the corporate managers reported that there was a specialist director for personnel and industrial relations on the main board. This was significantly less than earlier estimates (Marsh 1982; Millward and Stevens 1986). In 1992 the second company-level survey confirmed this level of main board directorship in large companies.

It also showed the use of the title 'human resource director' remained very rare at less than 10 per cent of the cases (Marginson *et al.* 1993). Interestingly, our case studies of nine multi-divisional companies give a strong impression of declining board membership. In 1984 four of the nine had a human resource specialist as a main board director. By 1987, in three of these four cases the personnel director had retired or moved and had not been replaced for policy reasons. In the five other cases where there was no personnel director on the main board; this had been a bone of contention for some time for reasons of status as well as because of the resulting perceived inablity to influence corporate strategy at the critical boardroom stage. In 1992 the position remained unchanged. Once a functional directorship goes it is unlikely to be re-established in the multi-divisional company, where the emphasis is placed on general management positions like divisional heads.

Both the 1985 and 1992 company-level surveys asked how important personnel and industrial factors had been in specific areas of business decisions regarded as being of strategic importance (Hickson *et al.* 1986). In 1985 respondents who said that personnel and industrial relations factors were taken into account in each specified decision were asked to rate this influence as compared with financial and technical considerations on a five-point scale, ranging from 'a lot more' to 'a lot less'. The rank order of subjects where personnel and industrial relations considerations were taken into account at the enterprise level was: plant closure (92%), redundancy (89%), run down of establishments (86%), production changes (82%), acquisitions (75%), investment (56%); and the setting of establishment budgets (55%). Personnel and industrial relations factors did appear to have been taken into account, especially where the decision had employment implications. When it came to fundamental first-order decisions such as acquisitions or takeovers, major production changes or fixed capital investment, personnel and industrial relations considerations were less likely to be taken into account and, where they were, were likely to have less influence than technical or financial factors.

The 1992 survey explored this issue in more depth by gaining the views of a finance executive as well as a senior personnel person. This allows us to compare different views of the role of the personnel function in strategic decisions. This is shown in Table

4.1. It is clear that personnel people were more likely to claim an involvement, especially in the vital stage of drawing up proposals, than their finance colleagues. It is often instructive to put data 'upside down'. So for example, in roughly half of the cases of mergers the personnel function was *not* involved in drawing up the proposal, according to the personnel executive who answered the questionnaire. However, according to the finance executive, the personnel function was not involved in two thirds of the cases. In the firms in the survey with decentralized industrial relation structures, where establishment managers were given only loose guidelines or complete autonomy, and where profit centres were set at the plant level (i.e. structures and control systems typical of financial control companies), there was a tendency for less attention to be given to personnel and industrial relations factors in these strategic decisions at the corporate level. In other words, there is evidence to suggest that the decentralization of operating decisions implied in financial economies is associated with a reduction in the strategic importance of personnel and human resources at the corporate level.

It might be expected that, where trade unions were recognized in every establishment of the enterprise, personnel and industrial relations considerations would be more likely to be taken into account in strategic decisions than elsewhere, because of union power. In fact the reverse was the case: there was a negative relationship between the influence of personnel considerations in strategy and union recognition. This was especially noticeable in 1985 between manual trade union recognition and the influence of personnel and industrial relations factors in decisions on takeovers, production changes, investments, and budgets. The reasons for this are not immediately apparent, but two hypotheses are worth further testing. Large, non-union firms, it is sometimes argued, appear to place greater emphasis on human resource management, and the survey data matches evidence from the USA (Kochan, McKersie, and Chalykoff 1986). This might of course be influenced by a desire to remain non-union. Secondly, the survey showed there to be a strong positive association between the size of the corporate personnel office and enterprise-wide union recognition, job evaluation, and corporate bargaining structures, especially for non-manual workers (Marginson *et al.* 1988: ch. 3).

TABLE 4.1. *Two perspectives on the role of the personnel function in strategic management (Percentage of Cases)*

	Merger/ acquisition		Invest in new location		Expand existing sites		Divest existing businesses		Closure of existing sites		Rundown of sites		Joint venture	
	(p)	(f)	(p)	(f)	(p)	(f)	(p)	(f)	(p)	(f)	(p)	(f)	(p)	(f)
Personnel function involved:														
in drawing up proposals	51	33	51	42	72	37	59	42	67	58	73	54	75	42
in taking final decisions	13	18	23	23	22	17	41	15	26	19	25	23	31	12
in implementation	64	55	66	40	48	42	67	42	56	50	51	39	50	41

(p) views of personnel respondent
(f) views of finance respondent
Source: Marginson *et al.* (1993).

The 1992 survey showed that this also applied to policies in areas like equal opportunities. Large departments are twice as likely to impose mandatory policies over the whole corporation as medium-sized to small ones, whose policies tend to be advisory. The strong impression is created that large, central personnel departments, especially in highly unionized firms, tend especially in highly unionized firms, to be isolated from strategic decision-making in the way identified by Nkomo (1988) and Christiansen (1987) in the USA. It is certainly the case that big personnel departments in the corporate office have *less* influence in business and corporate strategy formulation and in monitoring unit performance than smaller departments, who appear to work jointly with finance colleagues (Marginson *et al.* 1993).

The low level of involvement of personnel in fixed capital investment decisions and the setting of operating budgets for establishments in the 1985 survey is particularly noticeable. For each decision personnel and industrial relations factors were not considered at all in approximately 40 per cent of the cases, and where they were taken into account their influence was weaker than financial and technical factors, especially in unionized firms. Goold and Campbell noted in their study of sixteen diversified firms that the more companies moved towards financial control, the greater was the importance attached to the annual budgets, monthly reporting and to investment proposals initiated by the profit centre manager. Budget compliance became essential (1987: 129); annual budgets establish the key parameters within which labour costs must be managed.

In sum, there is reasonably strong evidence that employee relations and human resource issues are considered at the implementation phases of strategic decision-making where these have employment implications such as establishment run-down, closure, or disposal, and in exceptional circumstances of major industrial relations difficulties. There is much weaker evidence from the surveys of involvement in the formulation of first order strategies. The impression given by earlier studies of the weakness of personnel in this area is confirmed, especially that of the isolation of the personnel function and department from the strategic centre of many large companies. This is more likely to be the case where trade union recognition is widespread, and where second-order strategic decisions have led to decentralization and the establishment of local profit centres.

This lack of involvement in corporate strategy was clearly in evidence in seven out of nine of our case studies. It is important to note, however, that this does not necessarily mean that personnel and industrial relations factors exert little or no influence. Just as Hunt's study of mergers and acquisitions showed (1987), the failure to consider the employee relations implications of first- and second-order strategic decisions can have detrimental consequences. Problems with implementation may reduce the expected benefits of a strategic decision or occasionally, thwart or overwhelm the implementation of the strategy. It can also be the case that where those responsible for third-order employee relations strategies are divorced from first- and second-order decision-making, the policies they adopt may be unconnected, and may even clash with corporate strategy. For example, in two of our case studies the pursuit by some personnel managers of 'professional' standards of personnel management, implying to them integration, standardization, and co-ordination, was at odds with the first-order strategy of aggressive acquisition and the second-order strategy of decentralization, removal of the divisional tier, and business unit autonomy.

In the 1990s, with its agendas of delayering, flexibility, and performance, questions about the contribution personnel could and should make to help meet business objectives are often posed. The answer is that personnel staff are frequently placed in a residual role as implementors of other managers' strategic decisions. It is much easier to see the influence of corporate strategy on personnel than the other way around. It is very hard to find clear evidence of human resource questions actually deeply influencing corporate strategies and we suspect that many managers see no reason why they should. In one large corporation the group managing director noted that the board had decided on thirty priorities in the last few years, with the people ones being the most woolly, the hardest to measure, and the easiest to forget.

THE INFLUENCE OF CORPORATE STRATEGY ON HUMAN RESOURCE MANAGEMENT

The relative weakness of the personnel function and the lack of consideration given to human resource issues in corporate strategy should not surprise us. After all, virtually all textbooks on corporate strategy and business policy make no mention of employees, industrial relations, or personnel or human resource management, despite the current wave of interest in strategic human resource management.

The growing interest in resource-based strategies where attention is focused on the corporation's intangible strengths of people, reputation, and skills (Grant 1991) has opened a new and fruitful area for analysis of human resource management (Cappelli and Singh 1992). It is a welcome shift from the concern with market characteristics as the basis for competitive advantage. However, it still tends towards normative statements of the importance of people and skills without providing a guide to the type of actions or decisions that corporations can make to maximize their resources for competitive advantage, or suggestions on how to implant organizational routines and capabilities. It is hard enough to do this for a given business, but especially difficult to do it across the multi-divisional company. Some of these questions are considered in our final chapter on management style, where a means of analysing resource strategies is developed. Here, our purpose is to review the evidence of what actually happens before considering prescriptive solutions.

It is by no means easy to assess the influence of corporate strategy on employee relations. At one level it is clearly the case that the type of products the enterprise seeks to produce, the market it serves, and the technology it uses will influence unit size, geographical dispersion, and skill requirements, and that these will in turn influence, although not determine, the classic parameters of employee relations; in Gospel's terms work relations, employment relations, and industrial relations (Gospel 1992). More specifically there are, we believe, two features of first-order strategy (the long range goals and scope of the activities) and two features of second-order strategy (the structuring of the enterprise and the relationship between parts) which exert particular influence on management's approach to employee relations.

There appears, in fact, to be a small but growing body of research (Ahlstrand and Purcell 1988; Kochan, Katz and McKersie 1986) which has begun to suggest that most of the key (strategic) decisions made *within* the human resource sphere (what we refer to as third-order strategy) are, in fact, derivatives of first- or second-order strategy. Such research findings may not be pleasing to the personnel professional; not only (as we have seen in the previous section) does research show that personnel management has little influence on strategic planning; it now also appears to suggest that most of the significant or strategic decisions *within* human resource management are derivatives of higher-order strategic decisions. In the past, discussions on the rationale for choosing a given employee relations strategy or structure focused mainly on either the industrial relations logic of centralization/decentralization within the context of product market and production characteristics (Purcell and Sisson 1983) or on the basis of 'good practice' which all employers *ought* to follow irrespective of business circumstances. More recent research suggests a need to look directly to higher order decision-making.

The finding that key decision-making within employee relations is derivative from higher-order strategy holds, once again, for both sides of the Atlantic. In the USA, for example, Kochan, Katz, and McKersie have found that 'within most firms the motivation to actively pursue . . . alternatives was felt first by line managers and high level corporate executives, not by the industrial relations specialists' (1986:62). Within the UK case studies have shown that, where collective bargaining has been decentralized and corporation-wide internal labour markets decomposed, the prime motivation was the need to fit employee relations procedures and practices within second-order strategies as they evolved towards decentralization. We held discussions with ten large, multi-divisional companies which were planning to decentralize, or in the process of decentralizing or changing collective bargaining structures. Where the change involved a shift from multi-employer bargaining to corporate single employer bargaining this was strongly supported and championed by the senior corporate personnel staff. Where it involved a decentralization, corporate staff often resisted and discouraged the decentralization. This is explored in greater depth in Chapter 6.

CRITICAL CORPORATE HUMAN RESOURCE MANAGEMENT DECISIONS

How are first- and second-order strategies linked to third-order employee strategies? Before we explore this we need to provide a brief definition of third-order strategy within the context of the multi-divisional company so that evidence can be presented on the links. We define third-order or employee relations strategy in terms of four discrete 'critical choices':

1. the question of management style (whether or not a dominant style is cultivated across the entire corporation);
2. the deployment of personnel management resources and level of decision making within personnel and industrial relations;
3. the configuration of internal labour markets, seen especially in job grading and organizational layers and labour mobility; and
4. the choice of the bargaining and consultative units with trade unions and/or employee representative bodies like works councils.

These are key decisions which can only be taken at the corporate level and relate primarily to the definition of boundaries between parts of the company, both horizontally between divisions and strategic business units and vertically in the division of responsibility between levels of the firm. They do not in themselves constitute a complete human resource strategy since key decisions on job design, remuneration system, recruitment, and appraisal, for example, are very likely to be designed as part of the business strategies of divisions, SBUs, and operating companies. But it is these corporate decisions, taken in the light of the financial control system, which deeply influence the type of human resource management strategies chosen for purposes of competitive advantage at the business unit level.

In short, what degree of freedom is given to the local unit, and how does the enterprise manage the boundaries of its human resource system: through enterprise-wide systems, division-wide arrangements, or localized within the operating subsidiary, and to what extent is management within a given unit free to determine its own style and structures? To demonstrate the links between strategy levels we now consider how each of these key human resource management choices is linked to (and often determined by) higher order choices. We begin with the notion of management style.

Management Style

One aspect of first-order corporate strategy suggested by Andrews (1980) and others (Argenti 1974), is consideration of 'the kind of economic and human organisation [the firm] is and the contribution it makes, or intends to make to its shareholders, employees, customers and communities' (Andrews 1980: 18). Others refer to this as 'institutional strategy' (Hamermesh 1986) or more generally as organizational goals, values, or mission, describing the social goals and obligations of the firm. Within employee relations the term 'management style' is widely used (Purcell 1987). Is there a preferred way of managing employees? Are there limits and guidelines set? If a firm has an institutional strategy or a deliberate management style as part of its first-order strategies then, provided this is defined as anything beyond the banal, it will be liable to influence the conduct of employee relations, if only by distinguishing between acceptable and unacceptable management action. This is widely argued to be the case in firms like IBM, Hewlett Packard, and Marks and Spencer (although it is notable that detailed research has not been conducted in these firms to test these impressions, with the possible exception of Cressey et al. (1985: ch. 3)). But how extensive is the use of carefully considered management style set at the level of the corporate headquarters? Analysis and debate over the results of the 1985 survey (see Purcell 1987) lead to the view that 'by and large British management does not have a strategic approach to the management of people: pragmatism or opportunism continue to be very much the order of the day' (Sisson and Sullivan 1987: 429).

Our case studies of nine multi-divisional companies looked at the question of management style and concluded that none of the nine had what could constitute an effective, deliberate style. Two firms had attempted to define management style but in neither case was it enforced at all levels of the company (Ahlstrand and Purcell 1988: 8–9). Despite the widely held assumption that 'culture change' is both required and possible, the evidence for a lasting impact of such attempts to change management style is thin.

Hamermesh notes in his study of portfolio planning (commonly used in diversified enterprises) that firms excessively committed to portfolio planning techniques tend to ignore or find difficulty with 'institutional strategy'. This is related to the focus on the financial control mode of managing subsidiary operating units and the tendency to separate the enterprise into quasi-independent units where interrelationships are designed out or ignored, and central direction—administrative control—minimized. This is part of the second-order strategies that we noted earlier following Hill and Hoskisson (1987) and Goold and Campbell (1987); as firms grow and diversify, especially in conditions of market instability, emphasis comes to be placed on financial targets and accounting controls with less attention given to the human side of the enterprise or its place in the community. Thus the shape and size of the enterprise and the nature of the environmental conditions it faces are related to the type of second-order strategies that will be used and the relative importance attached to the management accountancy function compared with production and personnel management (Armstrong 1984).

The influence of 'the founding fathers' of the few companies known for their distinctive management style has been noted in both Britain (Purcell and Sisson 1983) and the US (Kochan, Katz, and McKersie 1986). What perhaps is of greater significance is that these firms have remained either vertically integrated or organized around the core business or critical function and, in the main, have grown organically rather than by acquisition. They have not developed as diversified multi-divisional companies. This is not to say that all integrated or synergistic firms have management style statements as part of their corporate strategy, nor that diversified, financial control companies cannot have an institutional strategy. But the inference, and, to a degree, the evidence is clear that the greater the degree of diversification and the greater the emphasis on financial economies the less likely, and the less easy it is for the enterprise to have a meaningful management style policy as part of its corporate strategy. In some cases, for example in Hanson, the question is not considered, while GEC are fond of saying that 'our system is to have no one system'. Style, beyond the requirement to manage the ratios, is the responsibility of operational management.

It is difficult for multi-divisional companies to decide on questions of management style, especially if they have grown, as is often the case, by merger and acquisition. This may be considered normal but it also often reduces the capacity of local units to determine what style they require for their business. This is explored further in Chapter 7. In the end, though, the question remains, as expressed by experts like Andrews, on the need for the corporation to be able to say what contribution it makes to employees, communities, share holders, customers, and suppliers—the stakeholders. These essentially moral questions are hard to answer but they do set the tone for what local units can and cannot do.

Corporate Personnel

Our case studies revealed that there was a growing uncertainty about what the role of the corporate personnel department ought to be. Research evidence regarding the link between the deployment of personnel (within and below the corporate office) and higher order strategy and structure is growing but far from definitive. Sisson and Scullion, for example, point to a significant amount of choice amongst companies regardless of strategy and structure (1985: 39) and we have earlier pointed to pitfalls in describing the deployment of personnel resources in terms of polar extremes of being either heavily decentralized or bureaucratized (Ahlstrand and Purcell 1988: 7–8).

Critically related to the notion of choice regarding deployment of personnel resources is the location of decision-making responsibility for personnel and industrial relations issues. One of the common characteristics of multi-divisional companies is the separation of strategic from operational management (Williamson 1970: 148). The decision to do this and the way it is done is part of second-order strategy on the structuring of the internal operating procedures. What is not obvious is whether industrial relations and personnel is defined as a strategic or operational management responsibility and whether the advent of human resource management makes any difference. If it is seen to be strategic, as it is in many Japanese corporations (for an example see Goold and Campbell 1987: 281–3), then centralized structures and administrative co-ordination managed by a central department or centralized function can be expected even if the firm is diversified. If it is defined as an operational matter then it is likely that operational or business unit managers will be given prime responsibility for the management of employee relations. There may well be head office monitoring and control just as there is in financial control, but the structures of employee relations will, as far as is possible, be designed to fit the structure of the firm. Where the structure is decentralized, moves toward human resource management will develop at unit level unless there is a strong overarching management style. But this is unusual in diversified corporations.

Hill and Pickering provide useful data from their survey of 144 companies, the great majority of which were multi-divisional (1986). Table 4.2 shows where responsibility for industrial relations and personnel decisions was placed according to their respondents. The modal position is for industrial relations to be defined as an operational responsibility. It was claimed that 62 per cent of operating subsidiaries were normally responsible for decisions in this area, compared with a quarter of divisional head offices and a tenth of corporate headquarters. The position for personnel decisions is similar but less marked in favour of operating subsidiaries. Of greater importance is the evidence provided by the authors on the link between levels of responsibility and profit margin. The location of decision-making was correlated with performance data: rate of return on sales and rate of return on capital employed. Hill and Pickering concluded that:

> Companies which allowed a stronger head office involvement in operating decisions tended to be less profitable. Similarly companies that involved their divisional head office in operating decisions also tended to be less profitable. These findings suggest that for optimum performance, operating functions should be decentralised down to the level of operating subsidiaries. This view receives support from the evidence of a positive relation between the responsibility of operating subsidiaries and profit in the case of buying, industrial relations and personnel decisions but, surprisingly, not in the case of production and marketing where no statistically significant relation was found. (1986: 47, emphasis added)

Looking closely at their data (table 16 p. 46) it is clear that the strongest association with rate of return on sales, loosely described as profit margin, of any of the variables was industrial

TABLE 4.2. *The location of decision-making responsiblity in personnel and industrial relations (%)*

	Company head office			Divisional head office			Operating subsidiary		
	Normally responsible	Shared responsible	Rarely responsible	Normally responsible	Shared responsible	Rarely responsible	Normally responsible	Shared responsible	Rarely responsible
Industrial relations	11	28	61	26	49	25	62	33	6
Personnel	16	42	43	17	69	14	43	50	7
No.	144			97			136		

Source: Hill and Pickering 1986(extracted from table 8).

relations, significant at the 1 per cent level for each of the three years tested (1978–80), with personnel next and sales only significant at the weaker 5 per cent level for only one of the three years. This clearly provides evidence that, from an economic point of view in diversified companies, industrial relations *ought* to be decentralized and made the responsibility of operating management.

One implication of Hill and Pickering's study is that it is decentralization and separation of businesses, not divisionalization *per se* which are key attributes of diversified firms. They note a growing tendency for enterprises to downgrade the divisional tier and 'further decentralisation of short-run decisions to subsidiaries within divisions' (1986: 35). This, we have noted, is associated with a change of strategic emphasis. The achievement of financial economies and the enhancement of financial control systems, the separation of business units, development of local profit centres, and the decomposition of divisions come to be emphasized. All these are likely to affect the management of employee relations as the institutions and procedures are forced to fit the new logic of the organization of the firm. The most significant element of the so-called 'new industrial relations' is the desire to make the management of employee relations subservient to corporate and operational needs. Management now shapes employee relations to fit its needs. It is in these circumstances not surprising that the term human resource management is preferred, since the pretence that employee relations is a 'neutral' activity does not apply.

The Structure of Collective Bargaining

The determination and control of collective bargaining also appears to be linked to higher order strategy. Numerous surveys and case studies suggest that some companies have simply not taken their bargaining structures as given but have forced through bargaining decentralization. Clearly there are complex reasons for shifts in bargaining levels. Five of our case-study companies had changed their bargaining levels and, without exception, their motivation derived from changes in the wider business structure/style and not from industrial relations imperatives. Structural changes included the creation of operating units as separate limited companies, while style changes included greater emphasis on 'entrepreneurship' and 'commercialism'. Principally, the stated aim has been to tie industrial relations and bargaining outcomes to the business performance of divisions or of the operating units.

The pressure to decentralize industrial relations was often resisted by the corporate personnel department and always opposed by the national trade union officers involved. In these cases, using the logic of industrial relations, there was often no good reason to change structures radically and personnel managers and union officers often emphasized the great risks associated with bringing about what they thought was an unnecessary change. Of our nine cases, one was non-union and another already decentralized (and acquired companies were expected to follow the decentralized route as soon as practical once rationalization had been completed). Each of the seven remaining cases attempted a form of decentralization with variations in type primarily related to differences in the previous pattern and the degree of production or operations integration between units. Such moves are now clearly impacting on the public service sector.

The incremental and political nature of strategy formulation was evident in five cases. As profit centres were established in operating subsidiaries, areas, or districts, and as budget compliance came to be more rigidly enforced (a typical by-product of financial control) so profit centre managers began to complain about head office overhead charges (especially those related to personnel), their inability to control labour costs, and the obligation to follow corporate administrative decisions unsuited to their precise needs. In this way they exerted a strong upward influence on strategy (Schilit and Paine 1987), adding to the downward pressure from corporate executives. This upward pressure could become more pronounced as unit managers were paid on an incremental basis according to the performance of their unit.

This upward and downward pressure often caused corporate personnel to respond by producing policy papers, getting consultants' or academic advice, attending conferences, or initiating covert discussions with national trade union officials in an attempt to respond to or resist the pressure to decentralize. Various compromise solutions were suggested or tried: 'tightly co-ordinating bargaining' (motor components); two-tier bargaining (transport services); 'enabling agreements' (privatized corporation); 'head office to attend all negotiations' (food); 'advance planning required plus enforcement of the corporate personnel manual' (leisure); 'only professionally trained negotiators can bargain' (bank). This defensiveness was understandable if only because the role of the corporate personnel department in its current form was at stake. In most of these cases the process of transition and adjustment continues, illustrating the emergent, incremental nature of strategic decision-making, or how strategy is made by implementation. In most cases bargaining is now fully decentralized, with the exception of transport services and the privatized corporation, where it has returned to a centralized system following the general failure of business decentralization policies in these companies in the recession of the 1990s and the effect of price regulation in a period of low inflation.

Internal Labour Markets

Choices around the configuration of labour markets have perhaps been the least explored of the strategic decisions in human resource management. We know little about why labour markets are structured as they are within multi-divisional firms. Why do some corporations have relatively homogeneous labour markets with career ladders which move across divisions and subsidiary companies, while others pursue a heterogeneous strategy with career ladders and structures specific to the division or operating unit? Clearly, there is a parallel here with the discussion regarding the deployment of personnel resources and the degree of diversification. Highly diversified corporations operating in distinct product markets may not see the same benefits from developing career ladders across operating units, preferring highly specific labour markets-linked to the particular product market. This logic would not stand, however, for more senior management positions where general managerial skill might be likely to override any product market linked skills. In any case, the most highly diversified firm, no matter how stunted its career ladders may be, must ultimately contend with the problem of management succession to the corporate office.

Hill and Pickering (1986) were dubious about the role of divisions or intermediary layers between operating companies and the corporate office. In four of our companies we were also puzzled

Corporate Strategy and the Influence of Personnel 73

by the role of the intermediary levels and concluded that in part their existence owed more to the need to provide a career ladder and jobs for valued managers than to a desire for efficiency. It is such jobs that have tended to go in the rush to delayer and minimize the steps in the hierarchy. This seems to have been the universal experience of all of our companies in recent years. A further trend has been to reduce the proportion of executives employed by the group or the corporation and instead to link management contracts with operating companies.

The process described by Hill and Hoskinson as the 'decomposition of divisions' is clearly taking place and in the process reducing the scope of the internal labour market across the corporation, cutting back on internal mobility, and changing the role and staffing of the corporate office in personnel; it is also linked to the decentralization of collective bargaining. In these circumstances it becomes more difficult to enforce or even articulate a corporate management style.

We are not suggesting that the explanation for decentralization can be found exclusively within first- and second-order strategies. A veritable host of environmental influences, from changes in labour law, the perceived decline in trade union power, the rise of individualism, economic stringency and product market competition, fashion, and the historical strength of workplace industrial relations, amongst other features, all contributed to the pressure first to question and then shift strategies and structures in employee relations and human resource management. However, in process terms, prime pressure came from corporate executives and line managers in their belief that it was necessary, and in the 1980s and 1990s possible, to bring employee relations into line with second-order strategies and structures as these became increasingly decentralized. Not to have done so, it was argued, could seriously have compromised the success of the decentralization to business units, reduced the economic gains expected, and demotivated unit managers who were told they were totally responsible for their units within the confines of the agreed budget. The development of M-form processes associated with diversification, divisionalization, and decentralization (and now divestment too) played a major part in questioning the logic of accepted patterns of industrial relations, while the decline in union power in the 1990s allowed corporations to realize their intentions without being stopped in their tracks by collective action.

These M-form processes are closely linked to the rate of change in corporate strategies and structures. The second company level industrial relations survey of 1992 asked the respondents what changes in corporate strategies had occurred in the previous five years (i.e. since 1987): 65 per cent had undertaken a merger or acquisition; 68 per cent had invested in new locations; 66 per cent had expanded on existing sites; 48 per cent had divested out of existing businesses; 62 per cent had closed existing sites; 40 per cent had run down existing sites; 35 per cent had entered joint ventures.

This rate of change was much greater than anticipated. Only three of the 176 large companies in the survey had not experienced any change. Even more interesting was the finding that 85 per cent of those companies engaged in a merger had also divested parts of their business in the same period (Marginson *et al.* 1993). Further analysis of the data dividing the sample between those companies which had engaged in an acquisition and those which remained 'organic' revealed that the former were much more likely to have decentralized structures held together by financial controls (Hubbard *et al.* 1993).

It is clear that the act of changing first- and second-order strategies requires corporate management to consider or reconsider third-order strategies. They have to take decisions in these four key areas of corporate human resource strategies: style, internal labour markets, relations with trade unions, and the role and staffing of personnel departments.

It has often been noted that the decentralization of employee relations within diversified companies does not imply unit autonomy (Kinnie 1987), although the degree of head office control has been disputed by factory managers themselves (Edwards 1987*a*). One limitation of all these studies is that the focus has been on control and monitoring of systems *within* personnel and industrial relations management. It is much more difficult to trace the effect of financial control systems themselves. The link with structure as firms come to adopt tight financial systems has already been noted in terms of decentralization. The final question is how financial control systems affect business unit managers themselves, and what influence this has on employee relations within the unit.

THE EFFECT OF FINANCIAL CONTROL SYSTEMS

Goold and Campbell note a tendency for the diversified firms in their study to have tightened the financial control over operating divisions while giving divisions or business units greater responsibility for their own business policy and planning. This was particularly noticeable in what they term 'financial control' companies (BTR, Ferranti, GEC, Hanson and Tarmac in their sample of sixteen companies). These companies:

> 'focus more on financial performance than competitive position. They expand their portfolios more through acquisitions than through growing market share. The style provides clear success criteria, timely reaction to events, and strong motivation at the business level resulting in strong profit performance. But it can cause risk aversion, reduce concern for underlying competitive advantage, and limit investment where the payoff is long term. Although financial performance in these companies has been excellent, with rapid share price growth, there has been less long term organic business building. (1987: 36)

There are three important attributes of these types of enterprises, which have often developed by acquisition into mature industries. First, 'there is no attempt to buffer the businesses from the requirements for short term profits . . . it goes much further than the capital market in applying tough standards' (ibid. 207). Secondly, 'these companies control only against annual targets and apply strict short-term (2–4 years) payback criteria to investment decisions' (ibid. 111). Thirdly 'they are willing to act speedily to exit from the businesses that are not performing or do not fit' (ibid. 126) and 'are quicker to replace managers, fiercer in applying pressure through the monitoring process and more effective in recognising and acclaiming good performance' (ibid. 132). Overall, the prime task in managing the business units (as opposed to marketing the products) is to reduce costs and push up margins in the short term. This comes about not simply because of intensified pressures in the product market but because of the way in which corporate headquarters set standards for unit profitability and decide on capital allocation, an approach which can lead to the social consequences of decisions being ignored (Mintzberg 1979: 424).

It is within this context that the growth of interest in employee involvement noted by Edwards (1987a) in his survey of factory managers must be set. Decentralized collective bargaining and worker involvement activities are aimed in part to gain employee compliance and identification with the business as a means of reducing costs, increasing productivity and effort, and eradicating indulgency patterns.

A worker might be offered 'involvement' but this depended upon accepting managerial definitions of goals and cooperating in their pursuit. A factory manager might have discretion but the outcome was carefully monitored and the expectation of high achievement was made abundantly clear. Devolving responsibility can create a more demanding environment than can imposing rigid rules. (Edwards 1987b)

Porter (1987) has noted how Hanson removed on average 25 per cent of labour costs from the companies acquired in their phenomenal growth in the last decade. It is quite understandable for unit managers to experiment with any relatively inexpensive means of securing worker compliance and co-operation, such as quality circles and direct communications, and to seek to boost performance and reduce relative labour costs, but they do so for opportunistic, short-term, cost saving reasons (Atkinson and Meager 1986), since the control systems under which they operate are based on short-term rates of return requirements. Cost minimization remains a prime objective which the performance control system reinforces. The contrast is with employee development policies where investment in people is emphasized. This also helps to explain the seemingly greater importance attached to the personnel management role at unit level in gaining employee involvement, if only to agree to the intensification of their own work. There is very little evidence generally that it heralds the widespread adoption of strategic human resource management systems (Guest: 1987).

If the pursuit of financial economies and the emphasis on portfolio planning systems tend to drive out questions of institutional strategy and management style at the corporate level, the use of stringent financial control systems in applying short-term performance targets for unit managers also tends to add an extra twist to opportunism, inhibit long-term investment in human resources, and focus on the short-term economic consequences of decisions (Mintzberg 1979). Opportunism is not necessarily explained as yet another instance of corporate myopia, however, since the short-term perspective of the stock market responds favourably to financial control companies with rapid share price growth, while diversified, long-range, strategic planning companies do less well in the capital market (Goold and Campbell 1987: 199) and are prey to take-over bids (Fortune 1988). In the light of this it is not surprising that emphasis is placed on financial performance and that operating units are set tough targets for rates of return on sales, capital employed, and output per head. In decentralization unit managers may be 'free' to do as they wish in human resource management, but financial requirements severely limit what is possible since all investment in people is a cost reflected in bottom-line performance.

Pressure is placed equally on unit labour costs and wage rates. Payroll costs nearly always form a separate entry in annual budgets which need approval from head office (Marginson *et al.* 1993). Most obviously the 1990s have been marked by a sustained effort to maximize productivity either through involvement, total quality management and quality circles, or by cost minimization seen most clearly in the reduction in head-count, the subcontracting out of services, and forms of flexible contract. This is the dilemma of human resource management: whether to emphasize cost minimization ('hard' policies), or 'softer' policies of maximizing the productivity of each person employed achieved by investing in people. The conclusion from the evidence we have is that the tendency for M-form companies, as they diversify, decentralize, and divisionalize, to adopt more stringent financial controls tends to reduce the possibility of human resource investment policies *unless* there is either an industry norm (i.e. everyone else does this for competitive reasons as in the motor industry) *or* there is a distinctive management style based around longer-term planning cycles. This is discussed in Chapter 7.

SUMMARY

In recent years considerable emphasis has been placed on the three levels of organizational analysis in industrial and employee relations: the strategic, the narrower policy level, and the workplace level (Kochan, Katz, and McKersie 1986). At the strategic level the debate has, for the most part, been restricted to the corporate response to shifts in product market conditions (Kochan and Chalykoff 1987), the effect of human resource management, and on the development of specific strategies towards the management of industrial relations such as, in the USA, avoiding unionization. In Britain the focus has tended to be on strategic choices in the structure and location of collective bargaining and innovative methods of segmenting the labour force and gaining employee compliance with management initiatives. The purpose of this chapter has been to provide an extra dimension to the study of the management of human resources by concentrating on the development of corporate strategies in large multidivisional companies and seeking to trace the effect of these strategies on employee relations management.

Our starting-point was the need to define the term strategy and especially corporate strategy. Rather than follow the classic strategy/structure distinction we preferred to focus on strategic decisions as those actions or plans which had long-term consequences for the behaviour of the firm. We distinguished between first-order strategies concerned with the long-term direction of the firm and the scope of its activities, and second-order strategies concerned with decisions on internal operating procedures used to manage parts of the business and link it with the corporate office. Here it was useful to see fundamental choices as being ones of integration or separation of the businesses, and of centralization or decentralization of profit responsibility.

We noted the tendency for large firms to diversify and as they did so, or as they grew in size, and when they faced turbulent product market conditions, for the control emphasis to shift from planning and the achievement of vertical or synergistic economies to achieving financial economies. Financial economies, implying an emphasis on the financial control system, were seen to have distinct implications for second-order strategies leading to the separation of businesses and the decentralization of decision-making. If, as is often the case, enterprises adopted portfolio planning methods to evaluate the place of each business in the portfolio, this also encouraged separation and the formation of distinctive business units. It is in this context that third-order strategies on employee relations were developed. These were seen to be largely subservient to shifts in first- and second-order strat-

Corporate Strategy and the Influence of Personnel

egy, although the possibility was indicated that in certain circumstances third-order strategies may be determined in isolation from first- and second-order decision-taking, something that can lead to difficulties of implementation.

We reached the following tentative conclusions. First, that in the formulation of corporate strategies in multi-divisional enterprises human resource management, and employee relations issues, are rarely taken into account unless the effect of a strategic decision has distinct industrial relations and personnel implications, seen for example in major redundancy decisions and plant run-down. Secondly, as firms shift towards the achievement of financial economies, the adoption of financial control systems, and the utilization of portfolio planning methods, it becomes more difficult for them to adopt institutional strategies or management style statements which provide the basis for coherent, corporate, standards of employee relations management based on beliefs about the best way to manage people at work. Diversity and opportunism are encouraged. Thirdly, the current trend towards decentralization in industrial relations, as seen for example in recent moves in collective bargaining in large multi-plant enterprises, is best explained not by strategic thinking by corporate employee relations executives, but by the need, sometimes the forced need, to fit the institutions and procedures of industrial relations into the mould created by changes in second-order strategy. That is to say that development in the structures of industrial relations in multi-divisional companies can better be explained by strategic choices taken outside the industrial relations arena than within it.

Our case study in the bank provided a fascinating counter-example. Here a newly appointed general manager (personnel) asked consultants to study the role and structure of the corporate personnel department. Armed with this report, which encouraged decentralization, he sought to develop a strong team of professional personnel managers in the key divisional posts, and to press for decentralization and greater line management responsibility and the adoption of modern human resource management policies. This ran counter to the centralist, traditional bank culture and he 'resigned' two years later. In this case he and one senior line manager (who also resigned) attempted to use change in third-order strategy to bring about wider structural change (second-order)—a process described appropriately by Legge (1978) as 'deviant innovation'. But such examples of strategic thinking in personnel are very rare. For the most part personnel is subservient to the needs of the prevailing culture of the firm and the traditions of welfare and service, taken with the blunted expectations of line managers of what personnel can contribute. What they are allowed to do limits their role in the management of change. In effect the 'logic' of the M-form company and processes ovewhelmed personnel. In the process their role in strategy formulation, while often dreamed of, remained marginal. Human resource management, whatever the rhetoric, is now seen as a third-order activity shaped greatly by bigger decisions of strategy and structure.

Finally, we hypothesized that the effect of financial control systems on unit managers, especially where the achievement of short-term targets was given primacy as it is in financial control companies, was to exert strong pressure on unit managers to push up margins, reduce costs, and limit investment to only those areas where there is a short-term payback. This in turn predicated a short-term opportunistic response by the unit manager with regard to employee relations (as opposed to the adoption of long-term strategic human resource management systems). The noted development in experimentation with means of achieving employee involvement was seen in the light of managers wishing to gain co-operation and acquiesence in cost-cutting measures and productivity improvements. The unit managers, especially in mature industries, were not so much responding directly to product market pressures or their place in the product market lifecycle, great though these pressures may have been, than meeting corporate targets designed to maximize short-term rates of return. This, it was noted, tends to reflect the priorities of the capital market in Britain and the USA. Corporate executives now spend far more time seeking to influence the financial markets and investment analysts than they do managing people, whatever they say about 'people being our most important resource' and despite their adopting the title of human resource managers.

There are, of course, exceptions, often quoted by the media. Great expectations are placed on green field site companies to break the mould, but the number of unambiguous and proven successes is so small that we are forced to ask not why good practice does not spread to other firms, but why and how these excellent firms exist in the first place. Often the answer is found in the origins of the company and the values implanted by the first owner and, more often than not, we find that their subsequent growth has occurred organically, not by acquisition, merger, or joint venture. Diversified corporations find it more difficult to manage the internal capital markets and budgetary control systems while simultaneously imposing distinctive long-term human resource management requirements on operating units. Most investment in people is a cost, and costs affect the bottom line, which becomes crucial in a recession. Operating managers confronted with corporate requirements to invest in employee development note that at the budget review time the bottom-line financial targets take precedence. And personnel managers are caught in the middle. They know the theory, perhaps, but have not the power to enact it.

This pessimistic conclusion can be tempered to a degree once personnel learn to think strategically and fit their policies to the corporate need, earning their place, so to speak, in the corridors of power. The next three chapters attempt to provide a set of conceptual frameworks to enable personnel and human resource executives to choose the best options and exercise strategic choice in the management of people in employment. Chapter 5 looks in detail at the role and organizaton of corporate personnel departments, Chapter 6 turns to the question of collective bargaining, and Chapter 7 concludes with an analysis of the difficult question of management style and the types of employee relations policies that might most appropriately be developed in different sorts of companies.

5

The Corporate Personnel Department

ORGANIZING THE PERSONNEL DEPARTMENT IN THE MULTI-DIVISIONAL COMPANY

Personnel specialists and personnel departments have difficult roles to play in any organization. The search for professionalism organized around a unique set of theories, models, and practices has never been firmly established if only because all managers have their own pet nostrums about motivation, selection, and the best way to manage people at work. Such problems do not, by and large, trouble accountants! It is also self-evidently the case that the personnel specialist can never have direct responsibility for managing 'personnel'—the employees in the firm. Whatever they do must to a great extent be filtered through line managers as the people with daily contact with their staff. Throughout modern industrial history the call has been to 'return the management of people to the line' where the responsibility, really belongs. It is better, however, to put this the other way round. By and large what personnel departments do is what they have been allowed to do by line managers and the board of directors. Ironically personnel shares this odd status with trade unions. It is often argued that if you want to know what trade unions are for you need to look at what they do and ignore what they say they do. Similarly, personnel textbooks describe an idealized version of the role of personnel. Just as with trade unions, what the personnel department actually does is very much a function of what it is asked and allowed to do. There is a further problem. It is extremely difficult to define the purpose of personnel management in a bottom-line sense. Since there is ambiguity over the purpose of personnel it is also hard to establish the best means by which these uncertain ends are to be achieved. It follows, as

Gowler and Legge (1981) noted, that it is inevitable that there should be no agreed criteria used to measure the effectiveness of either means or ends. Ambiguity is the hallmark of personnel; calling it human resource management changes little.

This ambiguity is, however, likely to be exacerbated in M-form companies since the vexed issue of centralization–decentralization adds an extra dimension to the expected tension between a 'professional' staff function interacting with line management. This extra tension is defined, for the most part, by the interactions that take place between personnel management at various levels and by the transactions of corporate personnel with line management at both divisional and plant levels. For this reason corporate personnel departments in multi-divisional companies will always have 'ill defined boundaries and muddy roles' as one of our respondents put it.

This was illustrated early on in the research programme by the experience of two case-study companies, 'leisure' and 'foods', in their attempt to find a role for corporate personnel. In the leisure company the corporate personnel department had never tried to bind divisions to a given management style or a single approach to personnel management. The lack of a distinctive style, or particular interest in one, was explained by the wide diversity of divisional product markets and histories and the lack of sustained interest by successive chief executives in questions of 'corporate culture'. The corporate personnel department had, however, tried to influence divisions by exercising administrative controls as embodied in the personnel manual. Fairly precise policy directives were issued covering such issues as union recognition, the composition of consultative meetings and remuneration policy. While the divisional personnel directors were personally committed to these policies and often remarked that they were indicative of the professionalism of personnel management, their line management colleagues were less convinced. For example, one policy concerned the need for all managers to be given appraisal interviews, but this was either not done or done too late in two divisions, because the divisional managing directors felt it was a waste of time.

Personnel directors in divisions were also required to produce annual and three-year personnel plans and pay strategy papers and get them agreed by their divisional boards. The plans were supposed to include timetables for the implementation of corporate personnel policies such as a reduced working week and harmonization. This proved difficult and some personnel directors were under pressure to avoid the issues by making only general statements, delaying the reply, or not submitting a paper at all. Significantly the harmonization plan was abandoned by the new chief executive. He saw little need for administrative controls and policies issued from the centre, preferring to allow divisions to manage their own affairs in accordance with market circumstances while conforming to strict budget requirements and the monthly reporting of results. He saw his role as exercising strong financial controls from the centre. Personnel was to him an operational responsibility. He would rather have had no corporate personnel department at all.

The ability of the corporate personnel department to influence management style or enforce common personnel policies was weakened by the process of decentralization and diversification. In the foods case study, divisional managing directors—the 'barons'—had become particularly powerful and each ran relatively successful business which made it difficult for the centre to intervene. Divisional personnel directors recognized their loyalty to the divisional managing directors and wanted only weak links with the corporate personnel department. It was generally seen as a sign of failure for divisional management to go to the centre for advice. Each wished to run its own activities as a separate business linked as loosely as possible to a 'friendly banker' at the centre and each was conscious of the strengths of its particular style of management. Corporate personnel was seen as a largely unnecessary and expensive overhead cost. Of course this view was not shared by the corporate personnel specialists, but given their lack of a power base and the dominant philosophy of divisional independence they could do little about it.

In the leisure case study, divisional personnel staff valued the help provided by the corporate staff, especially in codifying professional personnel practice in the manual. They felt that their first loyalty had, however, to be to their divisional managing directors who had little interest in following the administrative controls issued from the centre unless they had the clear backing of the chief executive. One personnel director noted that in terms of his own career it would be much more dangerous to disobey or cross swords with his boss, the divisional managing director, than 'fudge' or 'delay' the implementation of corporate personnel policies. Few, in any case, wanted to work in the corporate department since it was recognized as having responsibility without power.

The clearest indication of the weakening position of the corporate personnel departments in these two companies was the fact that both group personnel directors moved to new jobs during our research, but were not replaced. In 'foods', consultants were appointed to review the position in the light of organizational changes, while in 'leisure' significant reductions were made in the corporate office generally and it is felt unlikely that a corporate personnel director will be appointed again. This means that the only director, in both cases, to represent a specialist department or functional area on the main board was the finance director. This decline in the status of personnel management at corporate level was regretted in 'leisure' in particular since the divisional personnel staff felt that their profession had suffered a setback. They also concluded that the administrative controls and policies embodied in the personnel manual were clearly less important than the performance or financial controls managed by the finance function through the management accounting and information system. The new chief executive, it was noted, had previously been a finance director.

In 'foods' a group personnel director was eventually appointed in 1992, but his role and title (Human Resource Director) were markedly different from his predecessors. He has yet to gain the coveted seat on the board. Now all responsibility for employee relations is vested in the operating companies. The new corporate agenda is strongly focused on the management of managers—management development, succession planning, top executives pay for performance schemes, and the push into Europe with acquisitions, mergers, and joint ventures becoming the order of the day. This was reminiscent of the chief executive of an Australian company who told one of us in the late 1980s: 'I can no longer afford to employ two corporate personnel managers earning A$40,000 each in the head office. I can only afford one at A$80,000.' What he meant was that the administration of personnel and payment systems, the provision of legal advice, and managing relations with trade unions had to be decentralized. A new, more influential person was needed at the corporate level to work on a par with senior executives concerned primarily to improve the quality of management who in turn, in their operating companies, would be responsible for employee relations.

It is clear that the structure and configuration of personnel departments themselves appear to be deeply influenced by the wider organizational strategy and structure (Sisson and Scullion 1985). For example, it appears that functionally organized companies tend to operate with a different (more centralized) personnel structure than M-form companies. It seems, therefore, that personnel work in M-form companies is somehow unique and different, in both a 'management' (Purcell and Gray 1986) and 'structural' (Sisson and Scullion 1985) sense. It is for these reasons that we suggest that the study of personnel management requires sensitivity to differences in wider organizational strategies and structures.

The purpose of this chapter is twofold. First we attempt to work out how, in fact, personnel management is different within different organizational strategies and structures. Secondly, we examine strategic possibilities for the management of employee relations for the M-form itself. A variety of questions are raised and answered. For example, why do some firms have large central personnel departments while others operate with small central personnel departments or none at all? What are some of the recent trends in the structuring and ordering of the human resources department? What are some of the special problems, issues, and opportunities associated with structuring the human resources function in an M-form setting? Specifically: (1) where do you position the strength of your personnel department, at the central office, division, or operating company? (2) what should the role of the central personnel department ideally be (if you choose to operate with one)? (3) how do you manage a decentralized human resources structure (without losing centralized control)? (4) if there is no centralized corporate human resources department at all, how (if at all) are 'corporate' personnel issues worked out and resolved? These are the questions addressed in this chapter.

HOW BIG SHOULD THE CORPORATE PERSONNEL DEPARTMENT BE?

Why do some companies have vast central personnel departments while others operate with small central personnel departments or no central department at all? Is there some logic to the pattern of personnel structuring that we see in every day life? Sisson and Scullion (1985) have suggested there is and provide a convincing explanation. Briefly, they argue that the nature and extent of diversification influences the way in which personnel will be organized in a firm. They make a distinction between 'critical' function companies and those that have multi-based interests. They point out that those corporations that have more or less restricted their activities to a single business or range of products (e.g. Marks and Spencer, Ford) tend towards considerable integration of activities. In these businesses managers at the centre are not only responsible for developing business strategy but are also involved in a number of the 'critical' functions of operating management (Sisson and Scullion 1985: 37). As they put it:

Managers at the centre are responsible for a number of 'critical functions' of operating management as well as for strategic management. Personnel is one of these 'critical functions', because of the need for a standard approach without which the corporation would find difficulty in meeting its overall business objective'. (Sisson and Scullion 1985: 39)

According to the authors, the size of the corporate personnel office is itself linked to whether or not the firm is a critical-function firm or not. Single industry/product firms or those with one dominant market tend to have large corporate personnel departments, while in diversified corporations the need for variety leads to decentralized personnel structures. The pattern of personnel structuring (i.e. whether personnel is centralized or decentralized) is seen, for the most part, as a function of the degree of diversification.

Interestingly, Sisson and Scullion use the same kind of 'diversification' logic to explain differences in the structuring and organization of personnel resources which can be found within multi-divisional companies themselves. Here they have suggested that, other things being equal, the more diversified the activities of the corporation, the less likely it is to have a corporate personnel department. They suggest that companies like Cadbury Schweppes or Whitbread, for example, have middle-range corporate personnel departments because most of their subsidiary companies or divisions are involved in the same or very similar fields of activity. They point to other corporations, as for example Gallagher, which do not have a corporate personnel department at all because they are conglomerates with subsidiary companies and businesses scattered across a number of fields of activity (Sisson and Scullion 1985: 39).

The implications of the Sisson and Scullion thesis for organizing the personnel function are interesting, for it suggests that there is little opportunity to exercise strategic choice in the organization of personnel resources. It implies that the personnel structure that one operates from is for the most part constrained by the wider business strategy and structure. The design formula is, according to Sisson and Scullion, a simple one: if the firm has a low level of diversification and a narrow product focus then large, well-developed central personnel departments are preferable, while if there is a high level of diversification and a broad product focus then a small central personnel function with decentralized day-to-day personnel management is more likely. This model for design is a 'contingent' one: the personnel structure is contingent on the wider business strategy and structure.

Our own research generally supports the Sisson and Scullion thesis, but it also suggests that the 'agency of choice' is very much alive. In particular it reveals that the pattern of personnel structuring is just as much a product of whoever has the power to direct the organization. It shows, in fact, that firms can and do operate with a variety of personnel structures without incurring any serious diseconomies, that is, there is some degree of choice around the structuring and organization of the personnel function. In support of Sisson and Scullion, four of our seven case-study companies confirmed the diversification influence thesis. Three of our panel companies demonstrated, however, that the thesis was too limiting.

For example, our travel company went through a complete restructuring of its personnel function. Prior to reorganization, this firm fitted the Sisson and Scullion model perfectly. In accordance with its low levels of diversification and a narrow product focus the firm operated with a large, centralized personnel function. During the course of the research, however, a new chief executive officer was appointed, who was drawn from a company which was characterized by high levels of diversification and a complex product base. The new CEO imported 'decentralization' values into the travel company and, in a matter of months, exercised his power and dismantled and made redundant the entire central personnel function. All personnel matters are now handled exclusively at operating company level. While it was beyond the scope of the research to ascertain whether or not this restructuring had any effect on the performance and effectiveness of the company, there was no glaring or obvious evidence to suggest that this was the case either positively or negatively. In fact, local line and the personnel managers have without exception been pleased with the change since they gained power, authority, and prestige. In this case it was the values of the CEO which drove the rationalization. His view was that he was preparing the company for profit centre decentralization in order to diversify into other areas like retailing and hotels. The corporate personnel department, which had been highly successful in its own terms, got in the way of the push to decentralize and had to go.

There was other evidence to suggest that strategic choice is alive and well. Our retail firm, for instance, which exhibited low levels of diversification, had virtually no personnel presence at head office at all. Interestingly this company was also in the process of restructuring the personnel function in the direction of greater centralization. In this case, a new personnel director was appointed at the dominant or core business of the firm. The new director, coming from an American subsidiary operating in the UK, was an advocate of 'strong corporate cultures' and on the basis of this philosophy had been busy attempting to bring the group together through the development of common employment policies. His personal view of what the modern personnel department should do—the management of corporate cultures—led him to seek to centralize authority. Business failure in department stores reinforced this centralizing tendency.

Finally, in another case, we observed the acquisition of one highly diversified firm ('staple products') by another highly diversified firm; one of these firms, the acquiring company, operated with a small corporate personnel department (a total of three employees including a personnel director) while the acquired company operated with a large, well-developed corporate personnel department. In this case, the acquiring company 'exercised its choice' and made redundant the entire corporate personnel department of the acquired company within six months of the purchase date. The personnel director claimed after the restructuring that 'thousands of pounds had been saved in salaries' and the restructuring (i.e. redundancy of the central personnel department) had 'gone a long way in turning around an unprofitable business'. In fact no evaluation had been done on the role of personnel in the acquired company. There was a clear belief that centralization simply did not fit the style of the corporation.

Perhaps most revealing was the fact that in both 'travel' and 'staple products' the abrupt end of the centralized personnel departments was hardly noticed except by those who lost their jobs. It puts the classic theory of power into perspective:

> a department's/sub-unit's power may be said to reside in its ability to maintain its relative autonomy while increasing the dependence of other such units upon its capacity, exclusively, to fulfil their requirements. Hence, if a department is regarded as highly central (in the degree activities can affect those of other departments and the final outputs of the organization), if its coping activities are seen as both highly expert and non-substitutable, then, it is suggested, it will have power, due to other departments' dependency on it for the achievement of their own objectives and, at the individual level, for the rewards associated with such an achievement. (Legge 1978:27–8)

Choice in the structuring and organization of the personnel function was not simply between polar extremes (i.e. highly centralized or highly decentralized). For example, in one of our highly decentralized firms, the human resources function was not well developed at any of the three levels of central office, division, or plant. In this firm, our interviews began at the central personnel office, which consisted of two personnel employees. Expecting to find a more well-developed personnel function at either the divisional or plant level—we found none. We uncovered other cases which clouded the simple centralization/decentralization choice.

In two of our companies, the corporate personnel department existed in name only, with the personnel department of the dominant division or core business acting as the *real* corporate personnel department. The sheer size of the dominant division ensured that the 'corporate' policy was a product of the division rather than the central office (with pretences on both sides being maintained that the opposite was the case). This meant that corporate policy was policy which was produced either for the dominant business or by the dominant business, regardless of whether the policy fitted the rest of the group's companies and divisions. Pressures in the system were leading, however, towards a review of the current structure in both companies.

There is other data to suggest that the Sisson and Scullion theory is too rigid. For example, data from the 1985 company-level survey showed that the impact of corporation size, measured by number of employees, was the only statistically significant structural variable influencing the size of the corporate personnel department itself. The degree of product or service integration, the location of profit centres, and the type of employees seemed to have little influence (Marginson *et al.* 1988). This was replicated by the 1992 survey.

Interestingly however, the 1985 survey did point to other factors which seemed to be linked to the size of the central personnel department. The size of the corporate personnel department seemed to vary with the existence or non-existence of various personnel policies and practices. For example, the existence of corporate job evaluation machinery and joint consultative mechanisms were seen (perhaps not surprisingly) to be linked to large corporate personnel departments. Large personnel departments existed to service the institutions of collective bargaining and joint consultation, not to work on strategic issues of the development of the corporation. Not surprisingly, when bargaining was decentralized the *raison d'être* of such departments disappeared, and equally unsurprisingly they strongly resisted such moves out of vested interest.

One other fascinating finding from the 1985 survey and strongly reinforced by the 1992 results was the influence of ownership. Foreign-owned companies operating in Britain appeared to have a much more well-developed and considered personnel policy and department at both the corporate and workplace levels. Although foreign-owned companies were very similar to their UK counterparts in the extent to which they recognized trade unions and engaged in collective bargaining, there were marked differences, especially in the way they tackled corporate style statements and employee communication and consultation. For example, Table 5.1 shows the proportion of each type of firm which claimed to have a general policy or philosophy 'about the place of employees in the enterprise as a whole', as the questionnaire put it. We term this a management style statement. While most enterprises claimed to have a style statement, foreign-owned firms were more likely to commit this to paper and to give it to employees.

TABLE 5.1. *Management style statements: UK-owned and foreign-owned firms*

	% of enterprises	
	UK owned	Foreign owned
Style statement exists	83	90
Style statement written	45	68
Style statement written and given to employees	17	42
No. of enterprises	87	19

Souce: Purcell *et al.* (1987: 134).

Policies on specific areas of 'good practice' were tested in the survey. All but a small minority of both types of firm had enterprise policies on the provision of financial and other information, and few firms in either category had policies on job enrichment or autonomous work groups. Differences emerged toward policies on employee involvement. Foreign-owned firms were more likely to have policies toward quality circles and briefing groups but less likely to encourage profit-sharing and share ownership. Table 5.2 takes a closer look at policy and practice in areas of employee communication. This indicates that foreign-owned firms were more likely to have policies towards a variety of communications channels. The use of regular meetings between employees and senior managers, videos, suggestion schemes, newsletters, attitude surveys, and quality circles is particularly noticeable.

In terms of the type of information regularly provided to employees there was little difference between the two sets of establishments. Between 70 per cent and 80 per cent of both gave information on the financial prospects of the establishment.

TABLE 5.2. *Policy and practice in communicating with employees: UK-owned and foreign-owned firms*

	UK-owned	Foreign-owned
Junior management employee meetings	50	63
Senior management employee meetings	47	79
Systematic use of chains of communication	58	100
Videos	40	63
Suggestion schemes	26	47
Quality circles	15	37
Newsletters	68	100
Surveys/ballots	26	53
Other	16	21
No methods used	6	—
No answer/don't know	1	—
No. of enterprises/establishments	87	19

Source: Purcell *et al.* (1987: 134).

Millward and Stevens (1986) noted also that 'foreign-owned firms were noticeably more forthcoming with information about the position of the establishment and the enterprise as a whole but a higher proportion of foreign-owned firms also gave information on investment plans (70 per cent compared with 57 per cent of UK plants) and on capacity utilization (85 per cent compared with 66 per cent)'.

A higher proportion of foreign-owned firms were likely to monitor their establishments in terms of personnel and industrial relations performance. It also appeared that rather more use was made of the data collected. A fifth of foreign-owned firms said the data was used to aid investment decisions (UK firms: 10%). Just over two-thirds said it was used to draw comparisons with other establishments, compared with just under half of UK firms, and nearly three-quarters said the data was used to spot problems, compared with two-thirds of British firms. This was examined in another question on new technology. Three-quarters of foreign-owned firms said that the industrial relations record of an establishment was taken into account 'a lot' in deciding on new technology investment compared with just under half of UK firms. Perhaps this is an illustration of the changing influence of personnel management. Nearly two-thirds of the corporate-level respondents in foreign-owned firms claimed that the influence of personnel had increased in their organization over the last five years, and none claimed to have seen a diminution of influence. In contrast 40 per cent of UK firm respondents claimed an increase in personnel influence and 14 per cent said it had diminished. This is illustrated by the role of personnel in helping to draw up proposals for mergers or acquisitions referred to in the previous chapter. The 1992 company-level survey showed the personnel function was usually involved in this in foreign-owned firms (83%) whereas this was a minority experience in UK firms, especially UK multinationals (35%).

On average the foreign-owned firms in the 1985 sample had two-and-a-half times more personnel specialist managers employed in their UK headquarters than their British-owned counterparts. The second company-level survey of 1992 confirmed this difference and added a further twist. Figure 5.1 shows the existence of personnel directors on the main board and personnel policy committees by ownership. Overseas-owned firms are markedly more likely than UK domestic companies (i.e. those with no overseas operations) to take personnel seriously, as far as we can tell from structural factors (Marginson *et al.* 1993). It is interesting to note that this extensive use of personnel specialists is replicated at establishment level. On a much larger base of establishments, the 1990 workplace industrial relation survey noted that 'twice the proportion of foreign-owned establishments as indigenous employed designated personnel specialists' (Millward *et al.* 1992: 33). Thus there is clear evidence of substantially more resources being devoted to personnel and industrial relations management at both corporate and establishment levels in overseas-owned firms.

Two hypotheses can be suggested to account for this difference. First, foreign-owned firms are likely to be relative newcomers to Britain and were no doubt influenced by the image of British industrial relations that developed in the 1960s and 1970s. They were also less likely to be members of employers' associations and thus did not rely on external agencies for the management of industrial relations. Secondly, the act of investing in a foreign country and consciousness of culture differences means senior executives are less likely to have standard assumptions

The Corporate Personnel Department

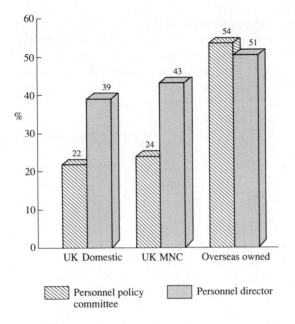

Fig. 5.1. Personnel Directors on the Main Board and Personnel Policy Committees by Ownership

about work and workers and the role of personnel. They tread more carefully and are more concerned to be accepted as a model employer, if only because of the assumed expectation that they will be different, either better or worse, but not the same.

Some evidence for this second hypothesis comes from Australia where a similar survey of the top 200 companies showed that in that country it was also true that foreign-owned firms were more likely to put a 'style' policy in writing and communicate it to employees than Australian-owned firms. However, within the sub-sample of foreign-owned firms it was the British who were the least likely to do so (33% compared with 57% of other foreign-owned companies) and the British response was very close to the Australian position (Deery and Purcell, 1989: 475). One has to conclude that a modern role for the corporate personnel office in British companies has yet to be fully discovered.

What has been the trend with respect to the way corporations are organizing their personnel function? Although the data is somewhat mixed we feel confident enough to suggest that a *decentralization* movement took hold in the latter part of the 1980s. Managers have begun to question the division of labour within the personnel function (in terms of the organizational level at which it ought to be developed) and have also questioned the role of the corporate personnel department itself. This soul-searching has, more often than not, resulted in the decentralization of at least some aspects of the personnel function (this has, for instance, often been expressed in the redundancy of personnel directors). Judging by the number of personnel directorships on the main UK boards of large private sector companies, the overall trend appears to be a decline in the importance of the central personnel function. Marsh concluded that, in 1980, 'few major multi-establishment companies employing more than 5,000 employees are now likely to be without a main board director with head office responsibility for employee relations' (1982: 62). This appears to have been modified by a 1984 survey which showed that 70 per cent of firms with 10,000 or more employees had a head office specialist director for personnel (Millward and Stevens 1986: 34). Data collected in the 1985 company-level survey painted a rather different picture. Here 31 per cent had a main board director wholly or mainly responsible for personnel and industrial relations. This rose to 34 per cent in enterprises with 10,000 or more employees. Differences in samples and levels of respondents make it difficult to compare the three surveys but the massive differences between the 1980 and 1985 surveys lends credence to the notion that the number of main board directorships has fallen in recent years. By 1992 the number of main board directorships seemed to have stabilized at 30 per cent. Put this another way, 70 per cent of the UK's largest companies do not have a full-time personnel or human resource executive on the board. This does make a difference. The 1992 results showed personnel people were more likely to be involved in first- and second-order strategy where there was a personnel director on the main board (Marginson *et al.* 1993).

The decentralization discerned in our main panel of companies confirmed this statistical picture. It was the result of change occurring both *within* and *outside* the human resource function. With respect to change *within* the personnel function we witnessed a gradual decentralization of key personnel activities downwards in the organization which in turn reduced the need for central staffing around these activities. For example, the recent trend toward decentralizing bargaining structures and the associated decentralized job evaluation systems have led to a drop in the need for corporate specialists in these areas. More recently the trend to decentralize responsibility (and more importantly budgets) for management training to divisions has reduced the number of head office management development and training specialists. This overall shift to decentralization of key activities has meant that corporate personnel people are playing less of an operating role within the sphere of personnel activities and more of a monitoring and control function. The emphasis here is placed on the co-ordination of divisional and/or plant personnel planned around a more broadly defined group policy or objective.

Not only did we discern pressures for change from within the personnel function in our panel of companies, but we also witnessed powerful pressure for change emanating from outside the human resources function itself. External driving forces for change related principally to change in the wider business strategy and structure. Without exception all of our companies which decentralized the personnel function did so as a result of a prior change to the business strategy and structure. The specific dynamics of this kind of induced change have been expounded in Chapter 3.

Why have we witnessed this 'personnel' decentralization wave in the latter part of the 1980s? Doubtless there are many complex reasons for this trend. First, we have seen a growth in the incidence of the M-form firm (and an associated shift away from the functionally organized firm). To the extent that the Sisson and Scullion thesis is correct we would expect that such a shift would lead to a decentralization of the personnel function (so that the function is better adapted to the wider structure). Secondly and relatedly we have also seen a shift in higher-level business strategies and structures (for example, the formation of autonomous profit centres). This shift has had a direct effect on the organization of central personnel as 'the role of the corporate office is often reappraised and itself often the subject of decentralization' (Marginson and Purcell 1986: 11). Purcell and Gray (1986, 220–1) have suggested, in this connection, that:

the greater the extent to which organizational structure and behaviour is altered . . . to encourage divisional freedom, the more likely it is that the corporate service units will lose authority. This will be the case where the corporate unit relies on administrative policies and controls more typically found in functionally integrated, bureaucratic organizations where rule conformity is the norm.

Thirdly, there has been a general trend in Britain in the 1980s to devolve key personnel activities (see, for example, Chapter 6 regarding the decentralization of collective bargaining). As a result, operational management and lower-level personnel departments have been strengthened to meet the increased demands placed on them to handle these new activities.

The 1992 company-level survey adds further, illuminating data based on the ratio of personnel managers in the head office per thousand employees. Overall the average is 1.8 per thousand. Not surprisingly this ratio declines the larger the corporation, with enterprises of over 10,000 employees having a ratio per thousand of 0.5. However, total size is not the only influence. Some 17 per cent of these large companies, each employing over 1,000 people in the UK, have a head office department with ten or more personnel managers. Here the ratio rises to 4 per thousand. These large departments are much more likely to be found when trade unions are recognized in all establishments of the firm. They are also much more likely than others to impose mandatory policies on gender and race equality across the corporation and operate a job evaluation scheme for senior managers. If we look at the data on the basis of company structure we find statistically significant relationships of some importance. For example, companies who have not embarked on mergers, have central strategic decision-making, are not divided into divisions, and are single businesses all have bigger central departments. There is clear evidence therefore that the process of divisionalization, diversification, and decentralization is strongly associated with the decline in size of the corporate personnel department (Marginson *et al.* 1993).

A further factor encouraging devolution has been the change in the external environment, especially the perceived decline in union power and the reduction in the volume of labour legislation that requires a corporate, as opposed to a union, response. It can be stated that the greater the pace of change in the external environment, particularly in relation to labour law, proposals for industrial democracy, and government intervention, the more likely it is that corporate personnel departments will be able to exercise influence, since a co-ordinated corporate response is required to what is seen as a hostile business environment engineered by the visible hand of government. In one sense, the strongest case for an authoritative, central personnel role exists when an interventionist government is in power, legislating across a wide area of employment-related issues, matched with powerful trade union leaders who have the ear of government. Thus, perversely, corporate personnel departments 'need' strong national unions and an interventionist, corporatist government to ensure a clear role in the management of human resources. This raises interesting questions for the 1990s with regard to the Social Chapter of the European Community.

It is all too easy for large central personnel offices to be relatively inert and disconnected from the wider business objectives. For example, we have argued (Purcell and Ahlstrand 1989) that while it may be taken for granted that large, well developed human resources functions are sensitive to the need to develop and maintain close links between business and human resources strategy, this type of department is in a prime position to use its status and power for self-serving behaviour to the point of actually blocking strategic change in order to protect its own institutional interests. The danger of leaving the management of human resources to large corporate personnel departments is that it segments personnel into an isolated department. Here the prime function is typically the management of relations with trade unions and the maintenance of agreed structures, as the 1992 survey data indicates. Personnel is thus reduced to an operational matter, even if it is located at head office, and is unconnected with strategic management. It has primarily a gatekeeper function: the act of divorcing trade unions from strategic management considerations is undertaken at the cost of personnel itself being excluded from, or being seen as having little contribution to make to, strategic management.

A decentralized personnel structure (with personnel responsibility devolved to the operating-unit level) does not have to mean a loss of control. Indeed, it may lead to increased control of human resources activities. A decentralized personnel structure simply means operating with different types of controls. Sisson and Scullion (1985: 39) point out, moreover, that,

> just because there is no corporate personnel department or it is very small does not necessarily mean that strategic personnel matters are being ignored. . . . Indeed, the logic of the 'multi-divisional' structure of organization is that issues are dealt with at the most appropriate level of the organization. In the case of personnel this means that they are dealt with at subsidiary company (or division) level so far as strategic matters are concerned and at the unit level so far as operational matters are concerned.

As they suggest, most of the large subsidiary companies or divisions of the 'top 100' corporations have fairly well-developed personnel departments 'the size and role of these personnel departments is comparable to those of some of the corporate personnel departments of the single business "critical" function corporations' (ibid.).

DETERMINING THE CORPORATE PERSONNEL ROLE

What then is the role of the corporate personnel department in the decentralized corporation? Its activities revolve around issues of role, corporate culture, dominant policies, interlocking boards and committees, crisis intervention, and surrogate financial control. Corporate culture statements (sometimes called 'motherhoods') are usually enshrined in general central personnel policies in the form of a brief set of statements or 'guiding' principles. Such statements are often developed at the informal level and are part of the underlying organizational culture, usually reflecting the values of the chief executive and the history of the corporation. They are not so much formal policies as philosophies. IBM and Tandy corporation, for example, stress the importance of individual growth and 'union-free' human resource management. This philosophy serves to set the tone for overall human resource management activities and has an impact on employee communication and consultative structures as well as employee appraisal systems. These broad philosophy statements 'encourage' each of the operating units to endorse certain human resources practices and not others. This is discussed in more depth in Chapter 7.

Control can also be maintained (more formally) in a decentralized personnel structure through the setting out of one or a few dominant employment policies, for example, through 'guideline or indicative' policies. These serve to trigger a debate about what is the appropriate policy and practice within each division. Guiding principles can be developed at corporate level which are loose enough to allow for flexibility and local adaptation and have a varied time-scale for achievement of minimum acceptable practice, yet are tight enough to ensure that local unit managers know the limits of what is acceptable and unacceptable.

A typical guiding principle is the designation of the corporation as an 'equal opportunity employer'. A simple corporation wide principle like this can serve to set the tone for managing human resources throughout the corporation. It 'informs' the subsidiary companies that this is not a 'cowboy' employer and that certain behavioural limits must be set in corporate operations. 'Foods' provided an excellent example of this type of corporate control through corporate culture or style statements. The need for rationalization in some parts of the corporation during the 1980s, the personal concern for employee relations matters of the chief executive, and the need to consider the human consequences of managing change, led to the development of a corporate policy document on managing change. This specified certain goals on such items as harmonization and referred to the need for full consultation with the workforce on the management of change, as well as the need to cushion the effect of change on individuals through retraining, relocation, and, where needed, compensation. The policy, as it worked out in practice, had three distinctive strengths. First, and most important, it clearly had the backing of the chief executive and the board; it was not just another corporate policy, nor simply the product of the corporate personnel department, but had special significance. While some divisions were less enthusiastic than others in following the guidelines, it was incumbent on each to show action or explanation for inaction. Secondly, rather than specify explicit goals, rules, or timetables for implementation, each division was asked to interpret the policy in line with its own needs and current styles of employee relations. Consultation can be undertaken in many different ways, from all-day meetings between directors and union officials to more informal contact between managers and employees in small non-union units. Thirdly, divisional boards were asked from time to time to report on progress made in implementing the policy, including the calling of special meetings of divisional managing directors and their personnel directors to reinforce the importance of the policy. Progress varied between divisions, and some managers in some parts of the business did not even know of the existence of the policy, but this was less important than the provision of standards of acceptable behaviour and style which influenced the way senior managers approached employee relations especially in their handling of change. Plant managers may not have known of the policy, but they did know, consciously or unconsciously, the preferred way of managing employee relations. They appreciated that their career did, to a degree, depend on how they conformed to the cultural norms.

Central policy statements are surprisingly easy to ignore and avoid in multi-divisional companies (and group executives often suffer from the delusion that a policy, once committed to paper, takes immediate effect), but acceptable normative standards tend to be pervasive if they are articulated and reinforced by the active involvement of the chief executive and divisional managing directors. Perhaps the most important aspect of such guideline or indicative policies is that they trigger a debate on what is the appropriate policy and practice within each division to justify why their style of employee relations is different from others but still appropriate for their circumstances. The power of such guideline policies is that they emphasize the achievement of consensus rather than relying on instruction. Thus large corporate personnel departments issuing mandatory policies on race and gender equality may well be less effective than those smaller departments providing advice and guidance.

This consensus-gaining approach is reinforced where there are 'interlocking board directorships'. Interlocking board directorships mean that the central office is never too far from any one operating unit: information can be exchanged on personnel policies within the various operating companies and 'problems' can be spotted before they get out of control. Our heavily decentralized staple products company made good use of this interlock system. The lone personnel manager at head office used divisional board reports as a key data base to monitor personnel activities around the company.

Periodic personnel committee meetings represent yet another effective way of managing decentralized personnel structures. All of our most decentralized companies operated with this kind of meeting, which can take many forms. They can be unstructured and *ad hoc*, or structured with formalized frequency. They can also take place at various levels of the corporation (for example, all personnel directors of each division can meet on a quarterly basis or all personnel managers of plants within a division can also meet). In terms of frequency, the most typical of our case-study companies had meetings for divisional personnel directors four times a year. These groups can either have a rotating chair or can be chaired by the group personnel director.

Sisson and Scullion have also suggested the tactic of group meetings:

Corporation-wide matters need not be neglected either. Even a single group or corporate executive can act as a convener for regular meetings of the personnel director and senior personnel managers from the subsidiaries to discuss matters of common interest. He or she can also be an important catalyst in the development of strategic thinking on personnel matters. For example, recent policy developments which, although taking place in the subsidiary companies or businesses, have had their origins in the corporate personnel department. (1985: 39)

There are still other ways and means of managing decentralized personnel structures. It is also possible to let the core business 'act' as the corporate personnel department. Here the corporation patterns itself on the personnel practices of the dominant or the core business. This may not be as unusual as one would think. Sisson and Scullion note, for example, that,

in the corporations without any personnel executives at headquarters, a similar role is often played by the personnel director of one of the larger subsidiary companies. Indeed, in a number of the corporations that have diversified from the core business—Gallaher and Security Services are good examples—the personnel department of that business fulfils many of the functions of a corporate personnel department. (Sisson and Scullion 1985: 39)

It is also possible to manage a decentralized personnel structure with the assistance of small central committees (composed of line managers and headed, perhaps by board members). This is one tactic many firms use to ensure that management development and career planning gets the attention it deserves. It is possible to argue, in fact, that many of the activities usually associated with central personnel departments 'can be adequately undertaken with a very small team of executives or even without a department at all' (Sisson and Scullion 1985: 39). Interestingly the 1985 company-level industrial relations survey found that the structural configuration of a corporate personnel function most closely linked with corporate strategy was the small number of cases where there was no corporate personnel department yet there was a main board director in personnel who operated through a personnel policy committee which included senior executives from other areas of the business (Marginson et al. 1988: 78). This is especially important in managing the most pressing human resource management issues of M-form companies—management development and career and succession planning. A decentralized personnel structure does not, therefore, mean a loss of control and resulting chaos. On the contrary, centralized control may actually be increased, albeit along a restricted range of control points. Control is simply maintained through a new variety of means.

As noted above, data, including our own case studies, reveals a trend toward the decentralization of the personnel function, evidenced as was suggested in the decline of personnel directorships. It is our belief, however, that there is danger in simply writing off the role and function of the central personnel department. What we need instead is a debate on and clearer consideration of the role of the corporate personnel department in today's business and economic climate. Data from our panel companies provides a starting-point for this debate.

Opinion was divided among the personnel specialists in the firms studied on the effect and importance of the change to a decentralized system. For some managers at division level the move was welcome, since it meant greater freedom to determine personnel policies in the light of the business needs of the unit and fewer niggling requirements to report to corporate headquarters. In their view, personnel practice could, and should, be closely tied to production, marketing, and operational management needs. Since the effective management team, charged with 'bottom-line' responsibility, was at division or unit level, the personnel function had to be exercised there, working with other executives and reporting to the divisional or subsidiary company managing director.

Others expressed regret at the decline of the corporate personnel department on two grounds. First, there was the loss of status; this was not important in itself (except in the implications for career opportunities), but it did imply the loss of voice and thus authority in the upper echelons of the corporation. There was a fear that strategic management decisions would now be taken without proper consideration of the personnel and manpower implications. These could be major issues, such as acquisition, plant closure and redundancy, or the allocation of capital expenditure, or seemingly minor technical decisions which are charged with emotional overtones. A sudden change in the company car policy was one such minor issue which caused a tremendous fuss since the decision came from the board without any involvement of personnel specialists.

On both major and minor issues the view was that decision-making was impaired at corporate level without the involvement of top-level personnel directors. More generally, personnel was now forced to be reactive, whereas what was required, according to some managers, was a proactive, strategically important personnel role. Second, and closely linked to the loss of 'voice', was the feeling that the abandonment of a co-ordinating authoritative corporate personnel department meant a weakening of professional standards, which had been replaced by '*ad hoc*-ery'. These managers argued that it was crucial to the longer-term development of their company as a business *and* social unit that attention should be paid, and seen to be paid, to human needs and aspirations, that resources should be allocated to allow personnel experimentation, organizational learning, and careful adaptation to environmental change.

Of course, these opposing views are not mutually exclusive, and each 'side' in the debate tended to exaggerate to make its point. Those who favoured divisional freedom tended to underplay the degree of control still exercised by corporate management through budgets and the monitoring of financial performance. It has often been observed that, paradoxically, moves to decentralize may often end up with the centre having even greater control. Policy guidelines are more easily broken and ignored than budgets and monthly reporting requirements.

Those unit managers in favour of greater autonomy for divisional personnel were often found in the bigger divisions and subsidiary companies, where quite sizeable personnel departments were in evidence, particularly if factory or workplace personnel staff were included in the count. The isolated personnel manager in a small unit or division was more likely to feel the loss of protection and companionship which the corporate staff had provided.

In the same vein, those who grieved at the loss of corporate personnel directors tended to forget that there still was a personnel presence in the corporate headquarters. The loss of a seat on the board labelled 'personnel director' did not mean the complete eradication of personnel management at corporate headquarters. Nor did the decline in the department mean that somehow division staff were unable to adopt a professional stance. Greater freedom often allows for experimentation unfettered by the need for corporate conformity. It was also clear that the corporate personnel role had not been as powerful as some who opposed the loss of director status implied. The loss was more symbolic than real.

If the need for a corporate personnel department is not as great as it appeared to be a decade ago, what justification is there for such a department now? When we looked at what was in the budget for such departments, some extraneous bits and pieces were sometimes included which often clouded the debate. For example in 'foods' there was resentment at what appeared to be the high overhead costs charged to divisions for central personnel services, but these charges included contributions to head office security, catering and medical services, corporate public relations, and community affairs programmes. In 'bank', payroll and pensions were included in personnel whereas they could just as well, if not better, have been placed in the finance function. These functions confirmed the impression of personnel as fulfilling an administrative role, such that few expected it to be policy orientated. Like all bureaucracies it was disliked: 'the department that always says no'!

NINE ROLES FOR CORPORATE PERSONNEL DEPARTMENTS

Once these and other peripheral activities are excluded, it is possible to identify at least nine 'core' activities which could form the heart of the corporate personnel department's role. They are outlined below, but this is not a prescription for all companies, since clearly much will depend on history, size, organization, and product market diversity, as well as the inclination of the chief executive.

1. Corporate Culture and Communications

Books such as *In Search of Excellence* (Peters and Waterman 1982) renewed the debate on corporate culture, focusing on the distinctive 'feel' that successful companies often have. They are bound together internally by more than financial resources and ownership; some organizations emphasize almost moral precepts or standards on how people, customers, and employees are to be treated, rewarded and encouraged, and, most of all, trained. While chief executives play a crucial role in formulating and enforcing such views, culture must by definition be more than the summation of one individual's philosophies. It often needs a Moses to write the commandments in tablets of stone and, in the secular world of business, to debate, articulate, and sell such philosophies if they are to be more than banal statements of the obvious or unattainable. One can hardly expect the finance director to do this, and other board members, representing divisions, are rarely able to take the long-term, global view necessary for such a task. If corporate philosophies are important it must be the personnel department's function to cultivate and disseminate them, or at least those which relate to the management of people. Guideline policies and long-term action plans which need to be adopted by divisional boards then follow, matched with multi-media corporate communication efforts which articulate these statements of vision. In recent years a debate about quality as in Total Quality Management programmes has put zest into the corporate culture debate and led to a rediscovery that styles in managing employees are important and do influence company performance.

2. Essential Policy Formulation and Monitoring

Some policy statements will derive from consideration of corporate philosophy. Others are more technical and detailed and usually found in the corporate personnel manual, if there is one. These may be broad or narrow in scope, including items such as employee relations matters. Other instructions might relate to items such as company cars, where equity is considered crucial and or where economies of scale can be achieved, as for example in the use of company training centres. Too often, however, corporate personnel manuals are ignored because they seek to impose an unnecessary degree of standardization. The key consideration is what is essential for all parts of the company or given levels of staff. If essential policy areas are identified, there must be some mechanism for monitoring compliance and identifying necessary deviations. Benchmarking within the corporation is increasingly used to audit key policies. Health and safety and the environment are good examples of areas where key policies are needed if only to protect the corporation from damaging publicity. Equal opportunities is another area where exhortations to be model employers are useless without monitoring, evidence, and audit but, as indicated earlier, this does not necessarily mean the imposition of rigid policies from the centre. What it does mean is that the corporation sets the standards in key areas and monitors the way in which divisions and operating companies set the policy designed for their needs. The key is the decision on which key areas require a corporate policy and statements of minimum acceptable standards.

3. Human Resource Planning in Strategic Management

The personnel and human resource implications of strategic management are often not considered adequately; one glance at the index of the classic textbooks shows that. We have been surprised to find, in our research, relatively few examples of serious human resource planning linked to strategic management. Yet personnel-type problems often emerge after a new company has been acquired or a major capital or divestment decision taken. These often relate to such issues as the failure to identify the drain on management resources in the parent company; the problem of integrating different cultures and practices; and thorny industrial relations issues such as pay parity or union recognition. Multinational expansion makes this even more important, partly because of legislation, but more crucially because of the need to manage and adapt to different cultures.

4. 'Cabinet Office' Services

Prime ministers in recent times have needed Cabinet office staff and 'think tanks' to provide direct advice and undertake specific investigations. Part of the reason for this is that departmental ministers, and especially civil servants, get drawn into defending their department and maximizing its interests, unconcerned about what happens elsewhere. Board composition in multi-divisional companies often resembles the Cabinet, with divisional bosses concerned primarily to protect their 'baronies'. If the only personnel presence is located at division level, reporting to divisional managing directors, the chief executive can be cut off and almost isolated. Big organizations are very political and it is not uncommon to find personnel directors playing a crucial personal role in working with the chief executive, especially in succession planning and identifying the next generation of senior executives, as well as undertaking specific studies requested by the board or in response to questions from non-executive directors. Without a corporate personnel presence, trusted and capable of carrying out this 'cabinet' role, chief executives can come to be neutralized and kept at arm's length by powerful divisions. As a result attempts to inculcate corporate philosophy and impose essential policies fail. By far the most influential personnel directors we met in the course of our research were those who had a close personal working relationship with the CEO. Their power was more personal than personnel: corridor power was more in evidence than functional authority.

5. Senior Management Development and Career Planning

This is often seen as the most important role and a number of companies who have taken the devolution route in recent years are concerned that they no longer seem able to plan management succession and development, let alone identify potential high-calibre managers. At the same time, given that in many companies senior managers have contracts of employment with the parent rather than subsidiary company there is a general need for a wide range of personnel services for this group of staff, from sophisticated remuneration packages to policies on alcohol abuse.

The use of assessment centres, psychometric testing and top-level strategic management seminars in leading business schools is indicative of a felt need to build management teams for the future. The more distinctive the corporate culture and the more specialist the firm, the more likely it is that senior executives will be grown from within rather than 'purchased' on the open market. This role cannot be undertaken properly without a corporate personnel presence of some standing. Indeed, while routine employee relations are devolved, management development has grown in importance, if only because of the growth in demand for the general manager to manage the proliferation of profit centres. It is here also via regular top management seminars that synergies can be discovered and nurtured between seemingly unrelated business divisions. If the pursuit of financial economies tends to encourage separation and local responsibility, corporate personnel, with the backing of the chief executive can help, in this way, to build integrative mechanisms which bind the corporation together and help the centre add value.

6. External Advocacy—Internal Advice

Who represents the corporation in the corridors of power in Whitehall and Brussels? And who joins the CBI and other national or industry-level bodies? Political lobbying and the assiduous pursuit of useful contacts in governments, trade unions, employers' bodies, city analysts, and among journalists is predicated on two principles: the need to influence and the need to know. 'Green and White papers' are an open invitation to give views, but much the most effective influence on policy-makers is exercised through informal contacts and explicit lobbying. Whatever the internal organization of large companies, civil servants expect the company to be able to express a corporate view on such matters and expect to have to deal with one person or department rather than a disparate group of divisional or business-unit managers, many of whom do not understand Whitehall, let alone Brussels. At the same time, the company needs advance information on proposed legislation and government policies in order to plan the necessary response.

One role for the external advocate is to keep the chief executive informed and instigate a debate inside the company on its likely response to proposals, for example those on the European Works Council. Corporate personnel staff have more time for such activities, and are often expected to inform and advise hard-pressed divisional staff, whose concerns are more immediate. Most important their power derives from an assiduously developed network of contacts. Again personal qualities are more important, than administrative expertise. This role grows in inverse proportion to the decline of employers' associations and trade associations, the traditional sources of advice and lobbying.

7. Information Co-ordination

The degree to which companies seek to co-ordinate the personnel policies and practices of operating companies and divisions varies widely, but virtually all companies have some type of co-ordination mechanism. Senior corporate executives therefore need access to effective information systems, which means the design of computerized personnel information systems has to be co-ordinated across divisions to ensure compatibility and determine common statistical yardsticks. More important still, some activities, such as pay bargaining, need to be co-ordinated to avoid internal conflict based on comparisons and to meet budget requirements. One mechanism for such co-ordination is the regular meeting of divisional personnel staff.

8. Internal Consultancy and Mediation Services

Organizational development (OD) has had a chequered history, but it remains true that intra-organizational conflicts can be tackled through a type of 'third party' role using process skills. Corporate personnel departments can emphasize this role since they are in a curious sense often almost neutral in intra- or inter-divisional conflicts and ought to have the necessary skills to intervene if requested or required. This same role can also be applied to industrial relations impasses if the corporate personnel department is prepared to get involved, although there is always the danger of encouraging unions to bypass lower-level management when they do. Much more important is an emphasis on organizational learning. Although much talked about, the 'learning company culture' fashionable in the early 1990s is easier imagined than discovered. Multi-divisional companies suffer from a 'not invented here' syndrome and find it extraordinarily difficult to translate good practice in one business unit into other areas. Green field sites have been disappointing in that good practice is diffused only slowly, if at all, through the corporation. Corporate personnel departments can play a crucial role here and some of the most innovative corporate videos have come out of their work in this area.

9. Personnel Services for Small Units

It is quite possible to envisage a situation in some companies where the number of full-time personnel staff is reduced at operating company level, to be replaced by a small central department which provides personnel resources as and when required. Even where emphasis is placed on reducing the central department by building up divisional staff, there are often units which are too small to justify their own full-time specialist. This might include the corporate head office itself. Arrangements in these cases are often made for personnel services to be provided on a contract or consultancy basis.

Two items often mentioned in addition to the list are training and pensions. Training, except for senior managers, is likely to be most effective when directly relevant to the business of subsidiary companies and divisions. The key corporate role in training is found in the design of annual or five-year rolling budgets. Does the corporate office require a training plan, or insist on say 2 per cent of turnover to be devoted to training? Unfortunately this is rare. Where it does happen it clearly forms a key part of the corporate culture statement. If the company is an integrated organization operating primarily in one main product market, then corporate training services are more important and relevant. Pensions are a more difficult area; a specialist department sometimes exists to provide the service, or else it is part of the responsibility of the finance department. Where employee-nominated trustees exist, the case for corporate personnel involvement is greater. Similarly some multi-divisional companies, but by no means all, retain some form of consultative committee or even bargaining at corporate level, which predicates the need for a corporate personnel department.

What is clear from the list of potential roles is that, as one manager said, 'corporate personnel departments have ill-defined boundaries and muddy roles'. Our research shows that the role and authority of corporate personnel departments is becoming yet more ambiguous and uncertain. Clearly there is a need for a corporate personnel *function*. Whether companies need *departments* of personnel professionals is another matter.

Much of the activity identified in the list of nine possible functions places a premium on political and interpersonal skills and 'corridor power'. In this situation the authority of corporate personnel staff comes more from their own expertise and style than from a clearly defined role and function. It has often been noted that personnel managers need to be adept at handling ambiguity. This is particularly true of corporate personnel staff, who must positively relish ambiguity if they are to survive.

One of the most frequently cited complaints lodged by line management at both divisional and plant level relates to the provision of central personnel services (this was, in fact, a chief concern raised by line managers against central personnel offices in our panel of companies). For many line managers there existed a discrepancy between costs charged to operating units for central personnel services and the actual service rendered. As one line manager put it, 'I see the bill but I don't see the service'. This was all the more evident for those operating units with their own personnel department and in some cases the complaint was fuelled by the local personnel department itself in search of greater autonomy and independence of action. For line managers, in these situations, personnel services were 'double dipping', charging out at both local and central levels.

There are ways out of this 'justification' problem (whether it is perceived or actual) and some firms have begun to experiment with innovative ways of providing central personnel services (not always, as we shall see, without dangers and problems of their own). In two of our panel of companies central personnel office staff have become redefined as 'internal consultants' and are now used solely on a transfer charge system. In this system, divisions and operating companies, rather than paying out a portion of the overall central personnel cost burden, 'pay' only for those services asked for and used. This might be quite a widespread practice. Just under half of the personnel respondents in the 1992 survey of corporate offices said the role of the corporate personnel department was to offer 'quasi-consultancy services'. We do not know what proportion of these are also required to charge for the service (Marginson et al. 1993). In this system, the group of internal consultants is interpreted simply as one of a range of consultants which divisions and plants can use if they wish. There are variations here and one of our firms employed at least one rider to encourage use of internal consultancy. In this case, the division and/or operating unit was compelled to give the internal consultancy unit an opportunity to bid for the job. Divisions or units were free to select from the range of consultants bidding for the job but had at least to have given central office itself an opportunity to bid. In addition, in this case, central office required a written justification if central office services were not used. This written justification was required, according to the central personnel office of this firm, in order to help them understand better the needs of divisions and plants (and not simply as a coercive device).

A variation on this theme, and perhaps more common, is the setting up of specific central office functions as independent subsidiary companies in their own right which sell services not only to the parent but to any other company wishing to make use of the service. This set up appears to lend itself best to the training and development function. This happened in one of our panel of companies. In this case, the training centre was turned around from running in the red (at approximately £1.5 million annually) to running at a profit (of approximately £300,000 per annum). The training centre, once used only for company employees, now provides training and development courses for small to medium-sized companies in the local area. In addition to ensure that the capacity of the training centre is fully utilized, it is rented out to other companies who wish to conduct their own training sessions and is available for hire for social functions such as wedding receptions, with priority given to employees. In another innovative example the training centre (relabelled the learning centre) trains child-minders for the local authority with the proviso that employees have privileged access to company-trained child-minders.

In one way, this 'internal consultancy' system can be seen as the ultimate test of the usefulness of the central personnel office in its relationship with divisions. At the same time it has the ability to resolve the vexed question of what specific services the central personnel office ought to perform. The market for services itself determines the activities and structure of the central personnel office. The system has also led to a much more fluid structuring of central personnel. As demand decreases for one service, that service can be dropped and replaced by activities with increased demand. It can be used, therefore, to test the need for the provision of certain central office activities. The system can also be used to test the competence and relative contribution of specific employees within the central office.

This structure is not, however, without its problems. Raybould (1985: 41) describes one firm's difficulties in moving to such a system and claims that the autonomy which was achieved by the operating companies under the new system did not make it easy to accept head office advice and that some of the central office staff were insufficiently trained to act as fully-fledged consultants. It is also limited to the provision of services to divisions and other parts of the corporation. As is shown in our list of nine roles, much of what the corporate personnel office does, however, is to serve the needs of the board and the CEO, and these key roles cannot be 'market tested' or sub-contracted. In some cases no overheads are charged out to divisions to emphasize this point, so that the department only plays a central role.

SUMMARY

This chapter has considered a variety of issues related to the division of labour within the personnel function in M-form companies. We have suggested that the overall trend in the 1980s and 1990s in Britain is towards a decentralization of the personnel function (down to divisional or plant level). British managers appear to be exercising 'strategic choice' in the way in which they organize and structure their personnel departments. We have identified both internal organizational and external environmental forces which account for this change. We have also provided examples and guidance to personnel managers in terms of how best to manage (tactically) personnel matters in a decentralized structure. Finally, we have suggested nine roles that corporate personnel can play. Ultimately, however, much depends on corporate culture and especially the values and beliefs of the chief executive and the roles that the board asks the department to perform working in collaboration with other senior executives in a personnel policy committee. It is here that foreign-owned companies are ranked as being more advanced than their British counterparts.

One thing is clear. Size is no substitute for influence. We concluded our study of the bank by noting how the well-staffed corporate personnel department, matched by regional and area departments, was a paper-driven system, not an influence system. Here, for example, all middle and senior job vacancies were internally notified to all divisional and area personnel managers. A massive trawl of appraisal and career planning forms was undertaken and the director with the vacancy to fill was presented with a dossier of the 'best' candidates for the post. Time and again the directors we spoke to said it was a waste of time. They knew all along who they wanted and they used their contacts to ensure they got them. Some organizations, especially in the public sector, go to extraordinary lengths to ensure that all forms of unethical discrimination in job appointments are removed, for example by taking away all evidence of gender, race, and age from application forms. The process is surrounded by rules and regulations. The aim is highly laudable but the bureaucratic nature of the regulatory process often actively encourages line managers to bypass the system. Rules do not change cultures but often reinforce them. Line managers in the bank also resented the paperwork associated with appraisals and would refer to the thickness of the corporate telephone directory to 'prove' how expensive personnel was. One divisional director estimated that 15 per cent of his costs came from central personnel overheads.

It was noted that large corporate personnel departments are closely involved in, and monitor, workplace activity. These were found in firms where the main institutions and procedures of industrial relations were usually located above the establishment at either corporate or division level, with tight instructions given to local management on what they could and could not decide in the human resource management area. These companies were also more likely than others to impose extensive controls of checking, information collection, and the maintenance of contacts with local management area. Unfortunately it was in these firms that personnel involvement and influence in strategic management was the least well developed. It would appear that the large corporate personnel department, which, for whatever reason, exists primarily to manage the institutional and regulatory framework of employee relations and personnel seen in bargaining, payment systems, and joint consultation, tends to become departmentalized and isolated from wider fundamental decisions in strategic management. As such, they function more as the managers of constraints, not opportunities. This is more likely to occur where union recognition is widespread. In contrast large non-union companies ironically appear to give greater attention to the personnel implications of strategic management and vest responsibility more in the line than in the department. It is these large departments that have most to lose in terms of their traditional role with bargaining decentralization, and most to gain if they discover a new, more dynamic role. The strategic issues involved in bargaining decentralization are considered in the next chapter.

6
Strategy and the Structure of Collective Bargaining

The structure of collective bargaining in a company is one of the most fundamental factors influencing the way relations with trade unions are handled and shaping the roles of personnel and line managers in dealing with industrial relations and directly with individual employees. A key aspect of industrial relations strategy has, in fact, been the choice of level for pay bargaining. Survey evidence indicates that few companies now take their bargaining structures as given. There has been a marked trend towards decentralization. This trend raises important questions about why companies are pursuing decentralization strategies, the appropriateness of such strategies for all companies, and the opportunities and problems associated with changing bargaining levels. Even though we are witnessing a movement towards the decentralization of bargaining structures there is remarkably little research on why a company decides on a particular level of bargaining and about the relative advantages/disadvantages of centralized and decentralized bargaining. Indeed, one personnel director we interviewed reported that in the process of trying to fill a senior industrial relations position he asked each interviewee to comment on the relative merits of centralized and decentralized bargaining. He claimed that not one interviewee was capable of entering into a reasonable debate on this issue and that he was unclear about this area; this uncertainty was one of the reasons why he asked the question in the first place!

SOME KEY DEFINITIONS

Before we take a closer look at recent trends in the organization of bargaining structures we need to set out some definition of terms.

The term 'bargaining structures' refers to the main institutions of formal bargaining over pay and/or conditions, and over the rules determining the operation of job grading, evaluation or bonus systems, and procedural rules. It does not mean the only place where bargaining occurs. Informal bargaining takes place to a greater or lesser extent in every system, but might be encouraged or positively required in some systems.

'Single-employer bargaining' is a loose term referring to bargaining which takes place within the company. One variant might be called corporate bargaining, where the whole enterprise is covered by one set of negotiations for a given group of workers, as in Ford or ICI before it was divided into two companies. Another variant is two-tier bargaining where local negotiations are allowed on incentive schemes or local additional payments. A third variant is divisional bargaining where each division of the multi-divisional company is responsible for the bargaining at that level for all its units. Some see this as the preferred method, avoiding the dangers of workplace bargaining, yet linked closely to meaningful product markets. A fourth and numerically the most important variant is decentralized bargaining, meaning mainly plant bargaining where the plant is one of many establishments owned by the firm. This does not mean sectional bargaining. Where a number of subsidiary companies operate on one site decentralized bargaining would imply bargaining at the level of each constituent company, as in GEC, and in this context would be different from site bargaining, which would cover all workers on the site irrespective of the subsidiary companies they worked for. Some companies allow for decentralized bargaining on pay while negotiating conditions centrally or via the employers' association. This is described here as split bargaining.

'Multi-employer bargaining' usually refers to employers' associations representing a number of firms bargaining at an industry, regional, or national level with the trade unions. It can also be undertaken via national joint industrial councils.

The very largest conglomerate companies may well have all types of bargaining taking place within the various divisions, constituent companies, industry groupings, establishments, and multi-company sites.

THE EVOLVING STRUCTURE OF COLLECTIVE BARGAINING IN THE UK

What has been happening to British bargaining structures in the last two decades? The single most important trend has been towards the decentralization of bargaining structures in both public and private sector companies. Multi-employer bargaining, conducted by employers' associations, is declining in importance in many industries like banking, retail distribution, and industrial cables while single employer bargaining is itself being restructured, often to the divisional or plant level.

With respect to multi-employer bargaining, the CBI pay data bank revealed that, by 1986, while 'industry-wide agreements continue to be of central importance in some industries, especially in the service sector, many larger firms which used to place partial reliance on them for non-pay matters have, as it were, come off the fence and brought the whole scope of their bargaining under their own control' (CBI 1988: 3). The CBI data bank also reveals that 'large firms have tended to decentralize their internal bargaining arrangements' (ibid. 5). Interestingly, the CBI also suggests that these same 'large firms have sought to reduce the complexity of employee groupings with which they bargain' and 'are showing a more strategic approach to the whole bargaining process' (ibid. 3).

The CBI's 1986 survey of bargaining structures is important to look at it as it is one of the few surveys which has directly addressed structural issues from a centralization/decentralization perspective. For example, their survey questionnaire asked the direct question about 'whether pay arrangements within the overall company had decentralized, centralized or remained unchanged over the last five years'. Table 6.1 provides the results of this question.

Over a quarter of the respondents said that their pay arrangements had become more decentralized, while a smaller proportion, 13 per cent, had moved in the other direction (CBI 1988: 33). The CBI survey also looked at bargaining trends by plant size. Table 6.2 reveals that decentralization was much more widespread in establishments with more than 500 employees, with well over a third of them reporting it (ibid. 33).

TABLE 6.1. *Decentralization and centralization pay arrangements over the previous five years, 1986*

	All establishments (%)	Establishments with collective bargaining (%)	Establishments without collective bargaining (%)
Substantially more decentralized	16	17	10
Slightly more decentralized	12	13	8
No change indicated	59	58	63
Slightly more centralized	11	11	11
Substantially more centralized	2	1	8
Total no. of establishments	346	283	63

Source: CBI 1988.

TABLE 6.2. *Decentralized and centralized bargaining: by plant size, 1986*

Size of Establishment No. of employees	Substantially more decentralized (%)	Slightly more decentralized (%)	No change indicated (%)	Slightly more centralized (%)	Substantially more centralized (%)	Total no. of co.s in sample
Less than 100	9	4	76	4	7	46
100–499	13	12	62	11	2	180
500–999	21	15	48	16	0	62
1,000+	22	17	47	12	2	58
ALL PLANTS	16	12	59	11	2	346

Source: CBI 1988.

The CBI survey confirms a diminution in the influence of multi-employer, industry-wide agreements and a pronounced growth in single employer bargaining at company or establishment level: 87 per cent of employees in plants with collective bargaining had their basic rates of pay determined at the level of the establishment or company (1988: 67). By 1986, therefore, single employer bargaining was dominant across the whole remuneration package.

Table 6.3 compares the results of both the 1979 and 1986 CBI surveys, looking at changes in the structure of bargaining with respect to the level at which rates of pay were determined. Figure 6.1 shows how the trend to single employer bargaining covered all of the main subjects of bargaining and the total remuneration package, and not just pay.

TABLE 6.3. *The matched sample, 1979–86: Changes in the structure of bargaining*

Level at which rates of pay are determined	No. of cases		
	1979	1986	1979–86 (% change)
Establishment	208	260	+25
Company	68	86	+26
National	48	46	–4
Multi-level	149	65	–56
Item not bargained	1	17	n/a
TOTAL 'MATCHED' SAMPLE	474	474	n/a

Source: CBI 1988.

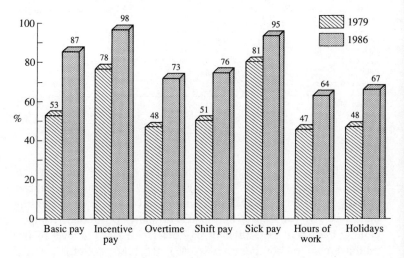

FIG. 6.1. Growth of Single–employer Bargaining in Manufacturing, 1978–86: % of Establishments with Single-employer Bargaining

Strategy and Collective Bargaining 123

The results indicate a pronounced move toward single employer bargaining. The extent of establishment and company-level bargaining increased by 24 per cent and 26 per cent respectively over the period in this matched sample of companies (CBI 1988: 35). Moreover, the number of companies who bargained at more than one level (shown as multi-level in Table 6.3) fell by 56 per cent. The CBI suggests that the majority are likely to have switched to plant or company bargaining only (ibid 35).

Interestingly, the CBI's 1986 survey of pay determination in private services reveals the same kind of decentralization trend but not to the same extent. Their results show a slight shift toward decentralization since 1979 (27 per cent of respondents reported that there had been at least a slight move toward decentralization). This is particularly marked in insurance companies. This trend to decentralization has been developing for a considerable period, but is associated with forms of corporate control and co-ordination. That is to say that it is still unusual for local bargainers in plants owned by larger enterprises to have a free hand in determining the bargaining agenda and outcomes. First, we have Marsh's survey (1982) of employee relations policy decision-making. Marsh's sample of British firms in 1980 revealed that over 80 per cent of individual establishments (which were part of multi-plant firms) saw themselves as determining pay and conditions (for both manual and non-manual employees) independently of their parent company. Head offices of multi-plant firms, however, saw themselves as determining pay for manual workers at all divisions and establishments in about a quarter of the cases, and again in almost a quarter of some divisions and establishments, with similar figures for non-manual workers. Marsh provides his own interesting interpretation of this data:

> It may be that establishments . . . are inclined to maintain that they have more independence from their head offices than their head offices would ascribe to them, or that head offices themselves may claim more authority than is their due . . . Some of those companies in the survey which recorded that they did not determine pay and conditions for all employees at company level appeared to exercise more direct regulation from that level than some of those which said that they did. (1982: 61)

It is evident that there can be few multi-establishment companies which have by this time failed to provide for co-ordination on pay and conditions claims of some kind, whether single or complex in character. No doubt this is one reason why such negotiations and their possible consequences are likely to be discussed by main boards more frequently than most other employee relations issues.

The comprehensive *British Workplace Industrial Relations 1980–84* (Millward and Stevens 1986) provides rather more detail. Table 6.4 shows the extent of consultation by local managements with head or divisional offices over their most recent establishment-level pay settlement. Millward and Stevens went on to note that: 'Previous research on plant bargaining in workplaces belonging to large organizations has suggested that the optimum arrangement is for local managers to have a high level of local autonomy within the framework of clear and agreed guidelines from the centre.' This point has been elaborated by Kinnie (1987).

TABLE 6.4. *Consultation by local managements with head or divisional offices over their most recent establishment-level pay settlement, 1984*

	Manual workers (private manufacturing) %	Manual workers (private services) %	Non-manual workers (private manufacturing) %
Consultation with Head Office[a]	66	83	61
Consultations after start of negotiations but before settlement	49	56	43
Level of consultations Head Office[b]	64	80	55
Consultations at both stages	39	55	37
No consultations at either stage	24	33	39

[a] Base is cases where consultations before start of negotiations are reported.
[b] Base is cases where consultations during negotiations were reported.
Base: establishments where managers reported that the most important level of bargaining over the most recent pay increase was at the establishment level.
Source: Millward and Stevens 1986.

Strategy and Collective Bargaining

The 'decentralization' trend shows few signs of abating. In fact, it is possible to argue that in a few years' time we might see much more local bargaining in central government, hospitals, and local authorities, as opt-out schools, trust hospitals, compulsory competitive tendering, and civil service agencies develop.

All the indications are that in the last half of the 1980s there was a substantial increase in the number of companies, especially large ones, withdrawing from multi-employer agreements. In some cases this led to the collapse of the employers' association or the termination of the national agreement. Notable examples are:

1. BICC, Delta Metal, and Pirelli General withdrew from the Joint Industrial Council for Electrical Cable Making on the same day in 1987.

2. Following the abolition of the Dock Labour Scheme, Associated British Ports devolved pay bargaining to the port level in 1988.

3. Coats Viyella, and Tootal, the major firms in the clothing and textile industry, withdrew from the British Clothing Industry Association and the British Textile Employers Association in 1988 and 1990 (see Leopold and Jackson, 1990).

4. Midland Bank withdrew from the Federation of London Clearing Banks in 1986 followed by three other major high street banks in 1988, thus ending one of the most effective national agreements in the service sector.

5. The ten largest supermarket companies in the Multiple Food Retailers Association withdrew from national bargaining in 1988.

6. Two independent television companies withdrew from the ITV Companies Agreement in 1988, leading to the collapse of the agreement in 1989.

7. A number of Joint Industrial Councils also collapsed in the period, for example in the cement industry.

Other areas where multi-employer bargaining ceased in the latter half of the 1980s were: newspaper distribution, provincial and national newspaper printing, the paint and varnish industry, the water industry, the steel corporation, and bus and coach industries. Two-tier bargaining with greater emphasis on local negotiations has developed in other areas, especially the meat and dairy industry (Industrial Relations Review and Report, 1989a, b, c, d, and 1990). The privatization of the electricity supply industry led to the complete abandonment of long-standing industry wide arrangements. By 1993 all negotiations over pay and conditions were undertaken at local level (ACAS 1993: 13). In the public sector the move towards a break-up of the national Whitley bargaining arrangements was slower in coming. None the less, the pressure for change in the Civil Service, local government, railways, and the National Health Service was only accommodated by the use of increasingly looser and more flexible national agreements to meet local needs and circumstances. Indeed, the emergence of Civil Service executive agencies with discretion on pay and grading, and the break-away of key local authorities from central bargaining highlighted the increasing trend towards fragmentation (Kessler 1990).

The strong growth of single-employer bargaining has been associated with decentralization to local unit level in many cases. This was frequently the case in those firms where corporate bargaining (covering the whole of the enterprise) had been established in the 1960s, sometimes in response to pay comparability pressures by union negotiators seeking to establish pattern bargaining. Examples of firms which decentralized corporate bargaining include the British Airports Authority, Cadbury, Lucas, Metal Box, Racal, United Biscuits, Massey Fergusson-Perkins Engines, and Pilkingtons. Elsewhere, in the insurance industry, national corporate bargaining arrangements came under pressure at Royal Insurance, Legal and General, and Eagle Star. In manufacturing, single-employer corporate bargaining is now very much the exception (outside the motor industry). In the service sector, corporate bargaining, or bargaining nationally within a division, is more common, reflecting the structure of an industry broken into small employment units (banks, shops, hotels) where decentralization to individual units would be more difficult (Purcell 1989). Decentralization of bargaining from corporate- to business-unit level was usually linked to the abandonment of corporate-wide job evaluation such that internal labour market structures were found within units of production, not across sites in multi-plant firms.

The 1980s witnessed the culmination of a process of collapse of multi-employer bargaining and simultaneous decentralization to local units. It was extremely rare for the trade unions to have initiated the restructuring of collective bargaining. In most cases they opposed it, unsuccessfully. It was management in large firms who took the initiative to develop firm-specific bargaining units within the enterprise and, within single employer bargaining, often to decentralize to the unit or divisional level. The government may have assisted the process through various legislative changes (especially the abolition of pay comparability mechanisms) and through deregulation in some areas such as the docks; but the key explanation for the direction and style of change came from broad strategic decisions taken by enterprises on their size, shape, and internal control procedures.

This picture was confirmed by the 1990 third workplace industrial relations survey (Millward and Stevens 1992), but this added an extra dimension, the growth of the non-union sector. In some cases decentralization is linked with the development of non-union plants, either because of derecognition (still a relatively rare occurrence), the opening of new green field sites where unions are not welcomed (this was the case in well over half of new sites according to the 1992 company-level survey) (Marginson et al. 1993), or simply because union membership withered away over time. This was the case in our electronics company where recognition was withdrawn when no one could be found to stand for election as an employee representative.

TABLE 6.5. *Basis for the most recent pay increase, 1980, 1984, and 1990 (%)*

	Manual employees			Non-manual employees		
	1980	1984	1990	1980	1984	1990
(a) Private manufacturing						
Result of collective bargaining	65	55	45	27	26	24
Multi-employer CB the most important level	27	22	16	5	5	7
(b) Private services						
Result of collective bargaining	34	38	31	28	30	26
Multi-employer CB the most important level	19	20	11	12	11	5

Base: establishments with employees named in column heads.
Source: Millward et al. (1992).

Table 6.5 uses data drawn from all the workplace industrial relations surveys (1980, 1984, and 1990) to show the marked fall in the proportion of establishments where pay was determined by collective bargaining and the percentage relying on multi-employer bargaining. The public sector has been excluded. The decline in the coverage of collective bargaining is very marked.

M-form companies, by virtue of their size and relatively large employment units, especially in manufacturing, are much more likely to recognize trade unions in some or all of their establishments. This was confirmed by the 1992 company level survey. Of the 176 large companies covered in the survey, 32 per cent recognized no trade unions, 25 per cent recognized them throughout the enterprise, 17 per cent recognized unions in most establishments, and 27 per cent recognized unions in some areas of the company (Marginson *et al.* 1993). Decentralization allows for patchy union recognition to develop and non-union areas to grow, since there is no longer a single corporation-wide policy requiring a standardized response across all parts of the enterprise.

In the USA similar strategic moves were focused on derecognition of trade unions (Kochan, Katz, and McKersie 1986). In the UK, as in most of Europe, the preference has been for reducing the disadvantages of unionism by weakening collective bargaining as the medium for the management of change, and on reliance on trade unions as the main link with the workforce. The development of single employer bargaining, often at a level consistent with profit centres or business units, enables the firm to bring in new payment systems and grade structures (as in banking) and link the management of labour more to the product market than the external labour market. It was this shift in emphasis, linking industrial relations with the needs of the business and away from external market structures, which was most obviously different and novel about the 1980s. These trends are much in evidence in Australia and New Zealand too as governments seek to deregulate national wage-fixing arrangements and encourage enterprise bargaining. What the Australians call 'enterprise agreements' are, in their large companies, most likely to be struck at a local level as in the UK, and in a growing number of cases no bargaining exists as the non-union sector emerges.

FORCES ENCOURAGING BARGAINING STRUCTURE CHANGE

Why is it that there has been so much change in bargaining structures in the latter part of the 1980s? Many factors appear important here. First, there has been a decline in employers' associations and employers' solidarity. British companies have begun to show a greater faith in 'going it alone' and a preference for more direct control over industrial relations outcomes. The pulling out of employers' associations may also be signalling a greater awareness of human resource management issues, of the need to internalize labour strategy (Gospel 1992), and to shape industrial relations strategy to firm and specific business strategies. It can also be said that reduced union power in the 1980s (resulting from legislation and labour market surpluses) has enabled British managers to make more a fundamental change to bargaining structures than would have been the case in the past.

It is also possible to identify a series of 'management style' changes amongst British managers which themselves appear to be encouraging thought about the appropriateness of centralized bargaining structures. We have already mentioned the increased interest of British personnel managers in making stronger links between business and human resources strategy. This has led to questioning about the value of multi-employer bargaining. There has also been a shift in emphasis towards more 'consultation' and less 'bargaining'. In an increasing number of companies, controls over bargaining have been matched by the development of consultative arrangements designed to encourage participation and co-operation, thus emphasizing the positive aspects of employee relations, especially at the local level. This shift in management philosophy appears to be encouraging the organization of smaller bargaining units which permit closer and more intimate communication.

There has also been an increased emphasis on the 'line manager' in the handling of human resources and industrial relations problems and issues. Line managers are increasingly being asked to take responsibility for industrial relations outcomes rather than relying on specialist industrial relations personnel. With the assignment of this responsibility, line managers have often demanded structural change in order that industrial relations responsibility should be matched with independence of action. Finally, there has been a rise of 'individualism' and a shift away from a more 'collectivist' orientation in many British firms. This 'individualist' orientation is reflected in moves to 'performance-related pay' (breaking away from the rate for the job) and the rise of the staff status employment contract (Ahlstrand 1990) and individual contracts, as in our privatized corporation, for managers who were previously heavily unionized. The term human resource management is used to signal the shift from the old order.

The linking of pay to productivity and productivity change can also be seen as a force leading to decentralization. Local initiatives needed to achieve high pay for high productivity are often more difficult to implement where there is corporate bargaining (as for example in the Post Office). Pay research tells us that productivity-related payment systems work best with individuals or small groups. Productivity bargaining, now often more appropriately called 'concession bargaining' itself implies local or plant-based bargaining. Decentralized bargaining allows for a more complex linking of pay and productivity at the local level, where the particular circumstances of the plant or unit can be taken into account and the knowledge of local managers and shop stewards utilized. Productivity bargaining requires an examination of working practices and systems with a view to determining more efficient working methods. This kind of bargaining can really only take place at plant level. Where industry bargaining or corporate wide agreements remain, these are increasingly of the 'enabling' variety, providing only minimum conditions and encouraging local additions or variations. This 'two-tier' type of bargaining is increasingly seen in the public sector and in integrated companies like British Telecom and ICI.

The key issue behind this new type of linked bargaining is the shift from concern with pay rates and levels of pay increases to fundamental issues of labour costs. Indeed in one of our companies the plan to decentralize bargaining structures was linked to a strategy where there would be a 20 per cent reduction in the number of employees, to pay for an increase in pay over a four-year period up to the level of the major competitor in order to recruit and retain better labour and a reduction in unit labour costs. The unit labour cost reduction was primarily achieved through the head count reduction (now called 'right sizing'!), changes in working practices, and the implementation of new technology requiring a different skill mix.

There have also been important changes taking place in the wider labour market specifically, a weakening of the 'going rate' and the beginnings of a regional pay philosophy. With the gradual erosion of the 'national rate' there is an associated pressure to adapt bargaining structures to capitalize on local and regional labour markets. Once again, this has meant a shift away from multi-employer and corporate-wide bargaining towards more decentralized forms which take advantage of geographical labour market variations.

In spite of the inherent logic of adapting bargaining outcomes to fit local labour markets, it is interesting to note that few of our own case-study companies which changed to decentralized bargaining did so to take advantage of 'cheap' local labour markets. This is in spite of the then 'north/south' divide and opportunities to take advantage of lower pay rates in the north. More than anything, there was company 'sensitivity' here, with firms wanting to downplay the 'north/south' issue. Indeed, in one case, a food company, local bargaining was used to drive up pay on the back of substantial improvements in productivity. The type of bargaining required at the local level is often complex, involving productivity or concession bargaining on flexibility and the utilization of labour.

The government has itself had a role in prompting change in bargaining structures. Notably, we have seen the end of many government controls—incomes policy, Schedule 11 and the Fair Wages Resolution, and, in 1993, wages councils. This has meant that the institutional mechanisms created earlier this century to promote fairness and labour peace, and to avoid undercutting and exploitation have been abandoned in order to free up the labour market. The consequence is that it is no longer possible to achieve 'flow ons', to use an Australian term, by resort to legal mechanisms. Changes in labour law have also restricted inter-plant comparisons and industrial action while emphasizing, through union ballots, local union democracy. Related to this, on the trade union side there is evidence of progressive decentralization with greater emphasis placed on shop stewards and district officers. Highly centralized unions, for example in rail, telecommunications, and the public sector, have had to develop local offices in response to decentralization by employers.

Many of these factors within industrial relations have tended to encourage bargaining decentralization. There are other forces in the realm of business strategy which have had a more powerful effect in decentralizing the bargaining structure. First of all it is clear that there has been a shift in both business structure and strategy, towards diversification and business decentralization, especially in the formation of profit centres. Here there has been an attempt to devolve accountability and responsibility and hence often industrial relations and bargaining activities down to profit centre level. Profit centre managers, having been given the responsibility for their own units, have sought to win control of bargaining processes, as they are tied to a key variable cost, the wage bill and labour productivity.

Our panel of companies revealed that the development of quasi-free-standing business units predated the move to bring bargaining structures into line with the new internal operating procedures. We encountered many cases in which profit centre managers complained that they could not control their labour costs if they were not allowed to manage and control their own collective bargaining. Some managers we spoke to complained bitterly that they were unable to manage effectively because they were not in control of their labour costs. They also complained about excessive head office overheads some of which were seen as being the result of an overstaffed personnel department responsible for collective bargaining. This in turn often put pressure on corporate personnel staff to include line managers in bargaining teams and then find ways of restructuring collective bargaining to fit the trend towards business unit separation and decentralization. In one of our case-study companies a new chief executive had arrived on the scene demanding decentralization of bargaining structures. The new CEO had worked for a conglomerate where the role of head office was financial management and industrial relations was defined as an operational responsibility. In this case the 'choice' to decentralize was forced unwillingly on personnel executives, whose experience suggested to them that there was nothing to be gained from decentralization (especially where the firm had integrated production or operation units and a tradition of central direction).

Without exception, motivations for decentralizing bargaining levels within our case-study companies derived from changes in the wider business structure or style, and not from industrial relations causes. The structural or strategy changes which promoted moves to decentralized forms of bargaining included the creation of operating units and divisions as separate limited companies, the development of more autonomous profit centres, and a shift in management style towards increased emphasis on 'entrepreneurship' and 'commercialism'. In virtually all of these cases, the principal stated aim was to tie bargaining outcomes more closely to the business performance of operating units and profit centres.

A change in the business strategy and structure did not, however, always lead to change in the bargaining structure. Indeed, we also witnessed resistance to bargaining structure change from within the industrial relations department itself. We referred earlier to the 'vested interest' pattern. In this pattern, employees within the industrial relations department are seen to have a vested interest in maintaining the current structure as their own status and power is derived from their own understanding of the system and how it works. For example, in the privatized corporation, the business structure was slowly evolving into a series of more independent profit centres. Local unit managers lobbied heavily for decentralized bargaining in order to gain control over what they saw to be a key cost area—their wage bill. This move to decentralization, however, was opposed by central personnel in a complex political battle involving divisional personnel teams as well. The central industrial relations department did not want to cede control of the centralized machinery as it was seen as critical to their own power-base in the organization. A major strike ironically provided the corporate personnel department with the opportunity to show that they alone had the contacts with union leaders and the bargaining experience to reach a settlement. It was in the union's interest to bolster the position of 'their friends' in the corporate office. The outcome was a return to centralization but this was unusual and is better explained by the nature of the firm and its integrated system, unique technology, and history than simply by power politics. It was probably a mistake to attempt to decentralize in the first place.

The key point is that bargaining decentralization is not a panacea, may not be appropriate in every case or even in most,

STRATEGIC CHOICE IN COLLECTIVE BARGAINING: CENTRALIZED V. DECENTRALIZED BARGAINING

The Parameters of Choice in Bargaining

The centralization/decentralization choice cannot simply be interpreted as a matter of polar extremes of either fully centralized or fully decentralized bargaining. For example, one of our case-study firms, the most avowedly 'decentralized', continued to operate a number of divisions within a 'multi-employer' bargaining structure. For this firm, multi-employer bargaining was seen to be compatible with an overall philosophy of decentralization, to the extent that it required minimal central office intervention and that local agreements needed to be struck to supplement the multi-employer umbrella agreements. This diversity in bargaining structures is not an uncommon situation. The largest of companies and conglomerates tend typically to have a mix of bargaining types throughout the company.

There is, in fact, an array of mixed and two-tier bargaining structures arrangements. In these 'mixed' bargaining structures, some items are negotiated centrally while others are devolved to the divisions or operating units. It is possible, even, to have 'multi-tier' bargaining structures in which bargaining takes place at each of the three levels of corporate office, division, and establishment. In this case, some agreements will apply to all employees of the corporation, others will apply only to divisional employees and still others to plant or establishment employees only. This bargaining arrangement may be more widespread than we think. We could well imagine, for example, a situation in which pension arrangements are negotiated centrally for all employees, with responsibility devolved to divisions for settling individual wage adjustments and operating units being given freedom by divisions to work-out individual productivity bonuses. In fact, all of our case companies revealed at least some form of pay determination taking place at each of three levels.

A study by ACAS in the early 1980s sought to examine the factors that appear to be associated with particular bargaining levels in the private sector. The evidence suggests, they reported, that industries which have highly competitive product markets, which are composed of a large number of small companies each with a small market share, and which are labour-intensive or geographically concentrated will, other things being equal, tend to have multi-employer bargaining. To some extent representational factors will work in the same direction, with both employers and trade unions in industries characterized by a large number of small firms and a competitive market preferring multi-employer bargaining. Single-employer bargaining is likely to be preferred by a company which is dominant in an industry, in industries where there are relatively few firms each of which has a sizeable market share and where companies wish to introduce their own wage payment system or productivity agreement. In many industries there are significant factors that influence employers and trade unions in both directions and what might be called organizational inertia may lead to existing bargaining patterns being retained, even though the reasons underlying the choice of a particular pattern have changed (ACAS 1983).

Where private sector employers have adopted single-employer bargaining, the available evidence about the factors encouraging centralized (company- or divisional-level) or decentralized (plant-level) bargaining suggests that economic factors are of significance. Decentralized bargaining is more likely in companies where average plant size is high, where there is significant variation in plant size, where plants are geographically dispersed, where technology and the product range varies between plants, where the wage payment system relates pay directly to effort, and where management is anxious to relate wage changes to productivity improvements. Examining representational factors, decentralized bargaining is more likely where management functions other than those concerning industrial relations are decentralized, where the philosophy of management is to decentralize decision-taking, where the industrial relations function is less influential in relation to other management activities, and where union organization is decentralized.

We have noted some of these factors earlier in this chapter, but the key issue remains to consider the advantages of a given bargaining structure and then to consider the factors companies need to examine in choosing a given bargaining level. In passing it is worth noting that the organizational inertia identified by ACAS in 1983 is now much less in evidence. The pace of change has quickened as markets have been internationalized and changes in organizational structures noted in Chapter 4 require firms to make choices. Both the third workplace industrial relations survey and the 1992 company-level survey note that the *least* change has occurred in the relatively sheltered and smaller UK domestic firms. It can be suggested that it is the experience of organizational change such as becoming multinational, opening new sites, or privatization which leads to the consideration of alternative methods of managing industrial relations.

ADVANTAGES OF DIFFERENT BARGAINING STRUCTURES

It is at those moments when choices have to be made that it is necessary to consider the advantages and disadvantages associated with different forms of bargaining. We will consider types of bargaining structure, each of which is open to M-form firms as a matter of strategic choice: (1) multi-employer bargaining; (2) centralized corporate bargaining; (3) decentralized bargaining; (4) two-tier bargaining; and (5) split bargaining. The location of bargaining at plant (fully decentralized), division, or company (fully centralized) level can be a key influence on the structure and behaviour of the unions in the company and indeed on the tone of industrial relations itself.

Multi-employer Bargaining

Multi-employer bargaining has the advantage of separating bargaining from decision-taking in a process which has been referred to as the process of 'institutional separation'. The extreme form of this type of bargaining has traditionally been found in Germany, where pay and conditions bargaining is conducted at the regional or national level between full-time union officers and employers' association officials. Multi-employer bargaining thus takes place outside the firm, leaving top managers and managers

Strategy and Collective Bargaining

at site level free to pursue co-operative and participative practices if they wish, with workplace representatives whose job is more consultative than concerned with engaging in adversarial bargaining. In this type of bargaining structure, the union negotiators are separated from both the workplace and top management. Institutional separation of the bargaining structures neutralizes union influence over decision-makers. Production managers may consult, but the right of last say rests largely with them.

The advantage of multi-employer forms of bargaining is that they focus formal negotiations in a given place and time and separate union negotiators from management decision-making both in the corporate office and the place of work. It is cost-effective since it reduces the number of bargains that need to be struck. Bargaining committees or councils act as buffers between a union's key activity (bargaining) and the use of techniques such as job evaluation and work study, to ensure that a distinction is made between the administration of agreements and the negotiation of changes in pay, procedures, and conditions. However, it crucially depends on employer solidarity and co-ordination not to break the agreed rate upwards or, especially, downwards and this in turn depends on each employer recognizing a collective interest in maintaining pay parity with competitors in the same industry. In sectors where there is growing import penetration this is very hard to achieve if imports compete at lower prices. Thus internationally the weakening of multi-employer bargaining is associated with the decline in tariff barriers and the growth in world trade. In our era of exceptional change it is also the case that multi-employer bargaining is ill equipped to be the forum for negotiating change. Indeed in Germany and Sweden there is evidence of moves to two-tier bargaining as changes in working practices and hours of work become necessary, and these can only be achieved through bargaining and debate at the workplace level. A further problem for M-form diversified companies is that they may straddle a number of industries so there is a need to be party to a number of settlements which may conflict.

If an M-form company decides to do its bargaining within its own walls, there are a variety of choices open to it. These can be developed along a continuum from fully centralized corporate bargaining (between bargaining units of like groups of employees across the entire corporation) to fully decentralized bargaining

(bargaining takes place at plant level, with each plant responsible for its own bargaining).

Centralized Corporate Bargaining

Centralized bargaining is preferred in some companies, especially those with integrated production units or national tariffs and standards of service to customers. It facilitates the possibility of the standardization of conditions in each plant and separates the union negotiators from the workplace. As bargaining is conducted outside the workplace, management is less likely to be encumbered with the need to handle pay conflicts and freer to pursue policies of co-operation and consultation. Change can be carefully planned and agreed and issues of fair pay differentials more easily achieved through an integrated corporate job evaluation scheme.

There are a number of other advantages associated with centralized bargaining arrangements. First, centralized bargaining serves to standardize employment conditions across the entire corporation. This may also indirectly serve a variety of other purposes, creating a more cohesive corporate culture and greater perceived equity amongst groups within the corporation, thus possibly reducing leapfrog pay claims. In very large companies corporate bargaining fills the role of multi-employer agreements, provides stability and rationality to the conduct of industrial relations, and establishes a mechanism for resolving disputes more effectively than decentralized bargaining. It also helps to avoid the trap that any change in local working arrangements, jobs, technology, and operating conditions will need to be paid for and bargained over. It is easier to develop effective joint working and consultation in the management of change if the adversarial bargaining can be kept out of the office or factory.

Others often argue the down-side risk that centralized bargaining inhibits local initiatives which attempt to link pay to performance and productivity. It could be argued that the push toward the multi-skilled, fully flexible worker could not have been achieved without local bargaining initiatives in the form of productivity-based bargaining. Finally, when companies work with centralized bargaining they are invariably constrained by the type and range of issues which can be brought to the table. Only those which are relevant to all parts of the company may be

Strategy and Collective Bargaining

introduced. It may be the case, for example, that only one operating unit is suffering from a shortage of skilled labour. With decentralized bargaining such a problem could be tackled through the bargaining process. Under centralized bargaining, however, management's hands are tied with respect to bargaining solutions to the problem and the temptation is to avoid or bypass the agreement or distort the job evaluation process with an artificial promotion of the job grade. It may also be that in certain labour markets the local plant is paying more than it needs to attract and retain employees. In short it may inhibit the achievement of flexibility to meet local circumstances.

Decentralized Bargaining

It has been suggested that decentralized bargaining is advantageous in that it denies the trade union any possibility of a role in corporate policy. It does so, very simply, by isolating the union from managers at corporate or divisional headquarters where top decisions are made and leaving the corporate or divisional office free to make the strategic decisions it wants. Strategic decisions, over acquisitions, mergers, diversification, or green field sites, never enter the bargaining agenda (simply because there is no union representation at the centre). The corporate office has no need to justify, let alone bargain over, these issues (which is why it is so alarmed at the prospect of the requirement to develop European Company Councils discussing these very things). The union's bargaining role is restricted to local issues and as such becomes extremely parochial. In 'staple products' the London head office prided itself that 'no union officer has ever crossed our portal'. The resistance of many companies to the formation of shop steward combine committees (i.e. multi-plant shop steward committees) is ample testimony of this. They do not wish to encourage co-ordination between companies in the portfolio. The 1992 company-level survey clearly showed that the greater the degree of diversification and the greater the use of financial controls the less likely were companies to allow such committees to exist. Interestingly a new development of the early 1990s was for the European Community to pay for employee representatives to hold meetings across Europe on a company by company basis. This occurred in the foods company.

This does not mean that corporate management gives local managers a free hand in bargaining; budget controls and centrally determined mandates provide control over bargaining outcomes. It is not accidental that annual pay awards often turn out to be much the same in all the plants of a multi-plant company, but the illusion of unfettered plant bargaining is maintained. Thus the focus on plant bargaining leaves broad company policy over such issues as investment or divestment unencumbered with negotiations. It leaves divisional and corporate management free to develop policy unbothered by the need to justify their decisions to a trade union, let alone bargain with them. It also enables management to argue that each plant must be productive and that each one's ability to pay must be the main consideration in setting terms and conditions of employment linking pay movements to productivity and change; that, at best, comparability must be restricted to the local labour market. Decentralized bargaining also 'fits' neatly with developments in the decentralization of management controls and with the formation of business and cost centres. It enables management to argue that pay needs to be linked to productivity and profitability of the individual unit or centre.

Even if bargained changes are agreed in corporate bargaining, there may be a lack of commitment at local level by managers and employees alike who were neither asked what changes were needed or possible, nor involved in the negotiation process. A classic case here was the deal a few years ago in the privatized corporation, where the union representing clerical workers 'agreed' to certain changes and got a higher than average award in return. Little change was achieved at local level and in some areas management had only a sketchy idea of the corporate deal, gave its implementation low priority, and took the view that the effort to turn promises into real productivity improvement was not worth it. If they had planned, costed, and negotiated the settlement themselves and worked directly with the employees affected it would have been more likely to be implemented in full.

Looking at those companies which have devolved their bargaining in recent years it is clear that for many it allowed them to bring in a number of other changes of a human resource management sort relating to payment systems, productivity, and the role of line managers. Indeed if the intention is simply to replicate traditional industrial relations patterns at a different level the change is not worth the effort. When our bank case study pulled out of the Clearing Banks Federation they changed the settlement date, introduced a new job evaluation scheme and took action to minimize the possibility of a legal challenge under the equal pay for work of equal value legislation. Line managers became more involved in the negotiation process and emphasis was placed on the role of staff representatives as union officers thus helping to develop an enterprise union structure less dependent on outside officials.

Decentralized bargaining should not be seen, however, as a panacea. It can potentially be destabilizing and runs the constant danger of inconsistency. The most frequently cited disadvantage of decentralized bargaining structures relates to the problem of managing the 'leap-frog' dynamic, in which bargaining across the corporation becomes nothing other than a 'referencing' activity or process. The centre needs to ensure that major changes in one unit will not create problems for another. One can appreciate the salience of this point when one realizes that it is often the case that different divisions and plants will bargain with the same full-time official as, for example, in the transport company. Indeed they gave up the 'pretence' of local bargaining in 1992 and returned to the centre.

In more practical terms, decentralized bargaining increases the actual number of negotiations and on a transactional cost basis is expensive. Do we really need all this bargaining? It also tends to run foul of local management inexperience in managing bargaining and negotiations. To decentralize bargaining requires not only a decentralization of the structure itself but also an associated decentralization of management expertise. There is a real danger that the result can be inflationary, since managers may follow the best local going rate but be unable to manage the compensating productivity improvements.

There are also problems associated with hostage-taking. As Marginson (1985: 4) puts it, 'weaker organized plants will hold back the stronger organized plants when it comes to placing effective pressure on management'. Thus, while decentralized bargaining isolates the union both from the power-holders in top management and from their colleagues in other plants, and limits the impact of strikes, company-level bargaining has the advantage of weakening union solidarity, making it more difficult to organize collective action (although once a strike begins it often lasts for longer than plant-based disputes). There are many variants between plant- and company-wide bargaining. We now consider two 'half-way' options: split pay bargaining and two-tier bargaining.

Split and Two-tier Bargaining

There are many variants between plant- and company-wide bargaining. Two merit further discussion. Some companies split pay bargaining at the workplace level from negotiations over conditions at either company or multi-employer levels. This type of bargaining tends to be favoured where there is a history of workplace bargaining and some form of inter-plant union co-operation such as a combined committee which is strongly defended by the unions. It is also favoured where enterprises have grown as a result of merger and take-over. Here the patterns of workplace industrial relations which developed under multi-employer bargaining are likely to be very different. The costs that would be involved in levelling up those plants with less favourable terms and conditions would in some cases be prohibitive and in any case might cause a reaction from those plants which have received more favourable treatment in the past. Far better in these cases to maintain the illusion of plant bargaining over pay, linked to the performance of the cost centre in profit and loss terms and to conditions in the local labour market, while focusing bargaining over conditions at company or industry level. Differences in pay levels between plants are much more easy to defend than variations in hours of work, holiday entitlement, sick pay, or pension rights, where union key bargaining tactics can more easily be mounted. The rapid spread of the shorter working week is a case in point. In split bargaining the 'corporate agreement' on hours and holidays, for example, plays the role of an industry agreement, establishing minima.

Two-tier bargaining is used in some enterprises, and this will increasingly be the case in public sector industries like the NHS, Royal Mail, and local authorities. Here basic terms and conditions are negotiated centrally but decentralized bargaining, often

Strategy and Collective Bargaining 143

at plant or area levels, is allowed for. This is especially the case where locally based incentive- or performance-related pay schemes are favoured. This has the advantage of allowing local flexibility to link pay to plant or regional performance and, to a degree, to local labour market conditions, and is particularly useful where the pace of change varies from district to district or where performance-related pay is favoured. Some two-tier bargains in fact give only little scope for local negotiations since the design of local payment systems and productivity bonuses is centrally determined. But it also needs careful control to ensure that bonus schemes do not decay and weaken the authority of the central negotiations and allow for 'coercive comparisons' to be drawn between units. The danger is that, without careful control of local bargaining, two-tier bargaining can be the worst of both worlds with inflationary pay settlements not achieving productivity improvements.

CHOOSING THE BEST BARGAINING STRUCTURE

The problem with these simple descriptions of types of bargaining structure is that no attention is given to the context of the firm. How does a company go about choosing the right bargaining structure? What factors should be considered? The case for reconsidering the structure of collective bargaining can easily be made, but it is no simple matter to choose the best structure for the needs of the firm and its industrial relations. A host of factors need to be taken into account in choosing the best format. Frequently the decision to change bargaining location has substantial knock-on effects on other factors in industrial relations and the organization of personnel and these too need to be planned for.

During the course of our research in companies planning for decentralization three main bunches of factors emerged as key influences in the choice of bargaining level. From this and earlier work in the area we developed a check-list of 35 points which need to be considered, sub-divided into three sets of factors which influence choice: corporate strategy and business organization; the configuration of labour markets; and structural and behaviour factors within industrial relations.

The first set of factors, and the most important, relates to corporate strategy and business organization. We have argued that the strategy and structure of the firm is the dominant factor in influencing collective bargaining and that it is the decision to develop the processes of M-form companies—diversification, divisionalization, and decentralization—which has had the most profound influence in large companies in recent years. However, this is more than fashion and choice, since the nature of the firm—the type of product and service produced is critical. It is extremely difficult to build a classic M-form company where there is an integrated production service supplied to a single market using a common technology.

For example, in 1991-2 the Royal Mail (letters) embarked on a process of divisionalization based on geographic areas. There was strong pressure from the new area boards to force through bargaining decentralization, in part to seek to break the national power of the main union. Postal Counters and Parcel Force had already been hived off into separate divisions. However, given the nature of the postal business it was extremely difficult to see how this bargaining decentralization could be achieved. The key requirement for boundary maintenance around each bargaining unit could not be met. Given this, the better option would be to recast national bargaining and allow for a two-tier structure if it was felt essential to bargain at local level on the substantial changes in working practices and staffing levels in the new customer-driven (and perhaps soon-to-be-privatized?) Royal Mail. Not all of these considerations apply in local authorities or hospitals, where there is therefore a greater chance of stand-alone units. Decentralization is not a panacea, does not apply in every case, can be expensive in transactional cost time, and can be inflationary. The key is the extent to which substantial advances in labour productivity can be achieved. If the new decentralized bargaining simply replaces previous agreements, practices, and patterns of industrial relations, then little has been gained and quite substantial additional costs incurred.

Using the check-list (Table 6.6), if most of the answers under the first section fall out on the right-hand side then we would argue that decentralized plant- or area-based bargaining is the wrong option. These will be companies which have a single or dominant business, have grown organically, have significant inte-

Strategy and Collective Bargaining

gration between production or service units, have a defensive first-order corporate strategy, and trade under a common logo with strong corporate culture.

The second set of factors refers to labour markets, both internal and external. The key issues here again centre on the extent to which the bargaining unit can be isolated from other parts of the firm, or whether it is desirable to do so. Clearly a geographic clustering of units in one town may make it harder to maintain the integrity of the bargaining unit, although there are some examples of sites being broken into separate bargaining units once operating companies have been established as profit centres, as in GEC Trafford Park, for example. Rather more important is the operation of internal labour markets across sites and the extent to which there is a unique set of skills in the firm not generally found outside it. Telephone engineers are a good example, and the existence of an integrated training centre for engineers makes separation much more difficult, as do the strategic business factors.

The factors in the third set, industrial relations, are much more malleable. Here the check-list is much more a means of establishing what other action needs to be taken if bargaining structures are moved than a list of determinants of the location. For example, where bargaining is currently undertaken at national or company level with national trade union officers, decentralization will require the identification, selection, and training of local negotiators in both management and the unions and it will be necessary for the company to help develop a domestic union organization. Similarly, great care has to be taken over the design of disputes procedures, job evaluation systems and information disclosure. Most important, local management have to have a clear idea of what they want to achieve once they become responsible for negotiation and consultation. They will need to look at their overall style of management; this is explored in the next chapter.

To use the check-list three questions have to be answered by those involved in the planning.

- Does the factor apply to my company?
- Does it point towards centralization or decentralization?
- What weight is to be applied to each factor? Not all factors will point in the same direction and a relative weighting,

such as scoring on a scale of 0–5, might be needed to indicate the balance of the argument for or against decentralization.

To reiterate: in each of the three factor groupings issues are arranged around a central question. An answer on the left indicates a preference or possibility of decentralized bargaining at plant, business unit, or subsidiary company level. An answer on the right shows a pressure toward centralized bargaining at major division or corporate level. Multi-employer bargaining is excluded, but answers which tend towards centralization could lead to multi-employer bargaining as the preferred option. For example, there was and remains a very good case for multi-employer bargaining to be developed in electricity supply covering Powergen, National Power, and Nuclear Electric, and possibly the distribution companies. This was not deemed attractive in the period of privatization since competition was the name of the game. But, of course, the attraction of multi-employer bargaining is that it takes wages out of competition while the real competitive edge of labour costs and labour productivity remains within management control. If a period of labour shortage returns, as it well might given demographic change, the attractiveness of co-ordinated multi-employer bargaining could be rediscovered.

Working with companies which are using this check-list it is evident that the knock-on effects of bargaining decentralization which caused the greatest difficulty concerned the restructuring of job evaluation, the redrafting of the disputes procedure to disallow automatic reference to the corporate office, and the consequences for the role and staffing of personnel. In one case the personnel director left, refusing 'to dismantle the system I have spent ten years developing', and in others head office staff have moved to local offices. Trade union opposition was also a point of concern especially among national officials who were devoted to national bargaining and feared a loss of authority. However, looking back at cases of bargaining decentralization the early fears have often proved exaggerated and the extra authority given to local managers often welcome. Sometimes a profit centre manager has been too enthusiastic in using the new freedom and has needed to be brought into line. Colleagues in other units and

TABLE 6.6. *Factors to Consider in Choosing to Alter Bargaining Structures*

Tending towards decentralization ⟵	Key question	⟶ Tending towards centralization
(a) Corporate strategy and business organization		
diversified	Are business activities . . .	single/integrated
by merger/acquisition	Has the company grown . . .	organically
no/little	Is the production integrated between plants/companies?	yes/a lot
aggressive growth	Is future strategy . . .	defensive/maintain market share
decentralized, local profit centres	Is accounting . . .	centralized
business unit responsibility, many brand names, no corporate logo	Is marketing . . .	strong corporate function, unified brand image, logo
varied, numerous, complex	Are product markets . . .	single, unified
locally designed, highly variable	Is technical change . . .	centrally designed, integrated systems
weak, performance-based	Is corporate culture . . .	strong, unified
considerable	How much discretion is traditionally given to local managers?	only a little
matrix, performance-based, business unit responsibility	Are management careers . . .	functionally organized, centrally directed

TABLE 6.6. Cont.

(b) Labour markets

local	← What is the market for key labour? →	national
yes	Are there significant variations in local labour-market conditions?	no
geographically dispersed	How dispersed are sites, business units?	concentrated
for only a few senior managers	Are internal labour markets a common feature in the company?	yes, for most grades of employee
no, most employees are recruited on the basis of previous training/skill attainment	Are skills unique to the company?	yes, most employees need significant, company-provided training
other workers in the community	Who do employees compare themselves with?	other employees in the company

(c) *Industrial relations factors*

NB: many of these factors will need to be changed with bargaining realignment

↓	→	
no, wide variety in recognition practices	Is the same union recognized throughout the company?	yes
no, unions are mainly general	Does the main union have a significant % of its members in the company?	yes, it is an industrial/company union
no, dispersed full-time officers	Does the main union have a tradition of head office control?	yes, head office based full-time officers
shop steward, lay officials	Who does the company prefer to bargain with?	national full-time officers
no, weak or non-existent	Is there a tradition of shop steward combined committees?	yes, they are well organized
restricted to local issues	In the past has information disclosure for collective bargaining been	centrally controlled, local profit figures not provided
often local, with bonuses, local rates	In the past has bargaining in the company been . . .	exclusively at corporate level
well-developed at local levels	Are consultative committees . . .	only at corporate level
no	Is there a history of interplant pay comparisons and co-ordinated industrial action?	yes

TABLE 6.6. Cont.

local level	⟵ ⟶	Where is the final stage of internal disputes procedures?	corporate/major division level
yes		Do fringe benefits and conditions of service vary between units?	no, common standards apply
yes		Are there wide variations in actual earnings and hours of work between units?	no, standard pay rates and hours of work apply
numerous and complex		Are payment systems . . .	simple/unified
locally based in business units		Is job evaluation . . .	company-wide
yes		Are there wide variations in labour productivity between plants or areas?	no
yes		Is there a perceived need for bonuses or incentive payments?	no
no		Is the corporate or major division personnel department well staffed?	yes
diffused and widespread		are negotiating skills . . .	concentrated

Strategy and Collective Bargaining

in headquarters often feared—or indeed felt—the consequences of a bargaining proposal which appeared out of line with current practice, for example a change in working hours or holiday entitlement. It was here that the co-ordination and planning role of head offices, through conferences and 'mandate bargaining' meetings, proved their worth.

In summary managers emphasized four main advantages gained in bargaining decentralization:

- enhanced roles for line managers;
- linking pay movements with productivity;
- close contact with employees, especially shop stewards and employee representatives;
- easier introduction of technical and organizational change.

The managers argued that the link with productivity and performance was most critical since it had affected labour costs much more directly than simple pay rates. Usually the preference has been to use redundancy and natural wastage as a means of reducing labour costs. None of our companies believed that marginally lower wages would have meant more jobs saved. It is also worth noting that two companies who thought that bargaining decentralization would be an obvious development concluded after careful study using the check-list, that it would be a mistake, especially given the integrated nature of their production or operations systems and centralized operating procedures. Bargaining decentralization is not a universal panacea, but it is worth considering, especially if pressures in corporate strategy are pointing towards business-unit separation, effective profit centres, and the growth of greater unit autonomy, or towards major changes in working practices.

THE POLITICS OF BARGAINING DECENTRALIZATION

Perhaps predictably, our panel of companies revealed that one of the greatest obstacles to changing bargaining structures came from within the personnel and industrial relations department itself. Our case studies often revealed the reluctance (and sometimes open hostility) of some industrial relations managers, especially at head office, to decentralize the structure of collective bargaining. A variety of 'rational' arguments were put forward by these industrial relations managers for sticking with the current system. Some claimed, for example, that there was no industrial relations advantage to be gained from the move since the existing bargaining structures had provided labour peace, competitive wage rates, and an effective mechanism for managing change. Others expressed fears that relationships with trade unions could be endangered, especially with national officers whose role could also potentially be radically changed and emasculated.

The Case of the Transport Company

For historical reasons deriving from its origins as part of central government, this profitable service company of 8,000 employees in eight locations determined pay in the early 1980s through corporate bargaining with a national job evaluation scheme providing an integrated pay structure for manual and staff employees. Each of the eight units had an identical structure and role and they varied only in size. The institutions of collective bargaining were focused at head office with extensive consultative and negotiating mechanisms established at that level to maintain the pay structure. Specialist personnel staff expert in job evaluation, employee relations, and training worked in the central office, running the employee relations systems and advising local line and personnel management on matters of implementation. The management style, as espoused by the chairman of the company, was one of extensive consultation and information-sharing, such that local subsidiary companies were expected to have consultative forums. There was extensive corporate information provision through company newspapers, employee share option schemes, and other media, all of which were designed to enhance employee identity with the corporation as a whole. A strategic business plan in 1986 (decided by the board without the involvement of personnel management), establishing local profit centres and greater responsibility for local management in running quasi-autonomous units, challenged the whole basis of employee relations. In future it was intended that pay should be related to local labour-market conditions through decentralized bargaining. Head office was to be reduced substantially and local managers free to recruit their own staff and determine their own style of managing labour. Company-wide job evaluation would be abandoned in favour of local schemes. In terms of our 35-point check-list, however, the case for decentralization was far from clear. The key to the plan was a diversification strategy into related businesses such as hotels, shopping centres, and engineering consultancy. This shift in first-order strategy led the board to set in place a second-order structure of decentralized divisions. Industrial relations as a third-order 'problem', was forced to bend to these stronger wishes.

There was considerable unease among trade union officers and the employee relations specialist managers at head office at these proposals. They argued that the centralized employee relations strategy based on stable institutional arrangements had served the corporation well, ensuring labour peace, competitive rates of pay, and steady technical change and modernization. Strategically, within employee relations, there was no reason to change the structure. The pressure to change came from a new chief executive with no experience in industrial relations and drawn from a highly decentralized company who loathed centralized systems, was adamant that there should be no personnel director on the main board, and believed that trade unions were anachronisms. To him the management of employee relations was the responsibility of local unit managers advised, if necessary, by local personnel specialists. The structure of employee relations was to be clearly subservient to the business plan of decentralization. He also challenged the prevailing culture of the corporation by announcing at the first group management meeting: 'Gentlemen [sic], we have three objectives: profits, profits and profits.' This challenged deeply held beliefs about safety and the standard of service. The outcome in this case took time to be worked through, and much depended on the power of vested interests in the trade unions and the central employee relations function to deflect the corporate plan, either overtly or covertly, and on the willingness of local management and employees to accept the challenge of local autonomy within the context of financial control.

In practice, after a serious strike threat, the firm allowed one key group of employees to have linked bargaining where the same deal would be applied across the board. For other employees decentralized unit bargaining worked for three years, with all local managers being trained in negotiating. However, the full logic of pay decentralization was not applied as the vital national union officers (to their irritation) attended all negotiations in each site. Not surprisingly the outcome was the same in each. When the firm hit financial difficulty in 1991 the chief executive left and the head office intervened to reach a deal with the unions, marking a visible return to centralization, much to the annoyance of local managers.

We would argue, however, that the 'rational' argument put forward by central personnel staff missed a key point, and was to some extent evoked merely to protect the status and authority of central personnel themselves. The driving force for change came not from the application of strategic thinking in industrial relations, but from shifts in the wider business policy. The need, as argued by line managers (and often local personnel staff who benefit from it), was to make the structures of industrial relations fit the corporate need of profit centre and business unit decentralization and help show a concern for profitability.

The Case of the Engineering Company

We can demonstrate some of the key issues involved in moving to decentralized bargaining through our engineering case study. This large engineering company ('motor components' in Table 1.1) with a corporate bargaining structure for its staff has been planning for decentralization for two years. The difficulty in making the decision revolved around three issues. How can you have decentralized bargaining where there is a corporation-wide job evaluation scheme? What do you do when the local bargainers fail to agree: should the centre intervene publicly under the terms of the existing disputes procedure? And what does local bargaining really mean? Should local managers in fact be free to do as they wish or do they need to be co-ordinated, constrained, or even told what settlement to reach?

The company's answers provide a useful guide to the difficult issues raised by decentralization. On job evaluation (a particularly contentious issue since it provides explicitly for pay comparisons between grades and within grades for people in like jobs). They concluded that there would need to be separate new schemes at each site and that the system would need to be sufficiently different to forestall pay comparisons. On impasse resolutions it was decided the centre would not get involved and the disputes procedure would terminate at local level. Each site must stand on its own feet. As one personnel manager in a different company put it: 'We consciously avoid meeting the union officials at a location other than where they are recognized.' The key issue in the engineering company then, was the need to build a 'boundary fence' around each site to avoid 'contamination', as they put it, leaking from one site to another. Ironically, just as the decentralized bargaining and restructured job evaluation was put in place to fit the new, slim, decentralized, divisionalized firm, customers began to complain about the fact that sales visits were made by managers from each operating unit, and began to pressurize the company for a more centralized, co-ordinated approach to sales and marketing. The pendulum effect between centralized and decentralized organizational structures was in evidence here. This uncertainty was reflected in the efforts made by corporate personnel to co-ordinate plant bargains while local negotiators tried to avoid the corporate gaze until after the deals were struck.

The key issue which led us to list the industrial relations factors which may need to be changed with a shift in bargaining level, is the need to build an informational and psychological boundary wall around the plant bargaining unit. The aim is to get employees and their unions to focus on plant issues and forget about other plants and any idea of collective solidarity. For example, in 'Foods' there was anxiety over employee reaction to the closure of a major plant. Would the other plants refuse to accept transferred products or machinery or even strike to support their colleagues who faced the dole queue? In practice, plant autonomy had by this stage grown so strong and the company had managed its communication on relative plant productivity so well (and paid above the odds in redundancy money) that employees in other plants guessed what was happening before the announcement and said, in the words of one convenor of shop stewards: 'They had it coming to them. Anyway, we are safer now they have been closed down.' The boundary wall had been established and national union officers were powerless to stop the closure. The point here was that the plants were in competition with each other. This had been encouraged through the use of extensive information on performance matched with decentralized bargaining. Union solidarity between plants had been broken. Had it been an integrated firm it would have been much more difficult to achieve.

A common tactic to achieve fully decentralized bargaining is to move first to various half-way houses or mixed or two-tier bargaining arrangements. One of our firms opted for this strategy, and, perhaps not surprisingly, this was the firm with the longest history of tightly controlled centralized bargaining. In this case, the choice to opt for two-tier bargaining was preconditioned by a past favourable experience with a centralized bargaining system. The two-tier bargaining structure gave a particular comfort to old industrial relations managers and trade unionists who had been 'born' into a centralized system. The tactic of leaving some items to be negotiated at the centre meant 'old timers' were able to be appeased, while at the same time significant decentralization was accomplished. Two-tier bargaining structures serve, therefore, as useful arrangements for warding off resistance to change while establishing the ground for full decentralization later. It is worth noting that we often found that organizational change occurred when a key individual retired or moved, emphasizing the short-term importance of personalities. It also points to the issue of strategic opportunity. There are certain times, like a retirement, when change is much easier to introduce.

One aim in managing decentralized bargaining is often to maintain a degree of control and co-ordination for the centre in spite of devolved authority and responsibility for bargaining. In doing so, firms try to have the best of both worlds: continued head office control and co-ordination accompanied by profit centre independence and autonomy. How is this accomplished?

Co-ordination of local bargaining is in fact commonplace, as the earlier data showed, and most companies with devolved structures either expect or actively organize co-ordinating activities. These include budget formulation and approval, personnel strategy papers, monitoring of labour costs, daily telephone contact when bargaining is in progress, and what we call 'mandate bargaining' meetings. Here senior personnel executives, and sometimes operating unit directors, meet to hammer out the bargaining strategy for the coming year. These can become very rough with managers arguing that a concession or tough proposal in one unit would have major repercussions on them. One manager commented after one of these meetings that it was a much tougher bargaining activity than he had ever experienced with the trade unions! Indeed he had learnt to bargain hard: seek to extract the maximum information from your 'opponent' while revealing as little as possible of your plans.

Co-ordination and control come first through the budgetary control mechanism, where labour cost targets are often specified in line with broader targets or rates of return on sales and capital employed. Secondly, it is common practice for co-ordination and monitoring activities to take place within the personnel function. Control can also be maintained through group personnel and the provision of key bargaining information. Through the systematic dissemination of information to the local bargaining units, each of the bargaining units can get an 'impression' of what settlement is acceptable (or not). This information can take the form of either industry settlement levels or of settlement levels within the corporation itself. One of our panel of companies actually used this kind of information in a tactical way and arranged bargaining information bulletins to reflect the desires of the corporation itself.

Divisions and plants face their own dilemmas in managing decentralized bargaining. Each division is anxious to preserve its bargaining freedom with the unions while, at the same time, protecting itself from dangerous or expensive precedents elsewhere.

Our first two case-study companies provided a rich source of data on these issues. The two companies studied, 'leisure' and 'foods', followed much of the trend towards decentralization discussed earlier. In the 1960s both belonged to the appropriate employers' associations and followed nationally-determined rates of pay for the various industries they operated in. By the mid-1970s both had to a large extent developed formal single-employer structures and in 'leisure' this led to a progressive withdrawal from the relevant employers' association. At the same time increasing pressure to devolve operating responsibility to divisional management, and an accelerated diversification programme, led to greater decentralization of bargaining units, with divisions free to determine the appropriate structures for their needs. Corporate bargaining in 'foods' was shifted to the divisions in the late 1970s. This was opposed by the trade unions and not welcomed by the corporate personnel department since it substantially changed their bargaining role. Only a residual national officers' consultative committee was retained at corporate level, but, in practice, corporate headquarters have not overruled divisional management by improving the final offer. In 'leisure', failures to agree in negotiations at domestic level could not be taken to the corporate level because of a clear yet unwritten policy on the part of corporate personnel not to intervene publicly.

The current structure of bargaining in these two companies is typically complex and diverse. In both, some divisions bargain at divisional level either because establishments are too small and numerous to sustain their own bargaining (leisure) or in two divisions (foods), because production planning (and profit centres) are determined at division level and the firm has traditionally bargained with full-time union officials. Elsewhere plant negotiations are preferred. In some cases this is because of a mixture of historical precedent derived from the engineering industry and emphasis on establishment-based profit centres (leisure). In the new products division of the foods company profit centres are similarly decentralized, but a further influence has been the divisional managing director's desire to distance his division from that of the parent company. The more decentralized the bargaining the less visible are the process and outcome of the bargaining to corporate headquarters and other divisions. Decentralized bargaining also fitted his own preference of dealing directly with employees while avoiding outside union interference (i.e. full-time officials).

Some degree of co-ordination was attempted by corporate headquarters over bargaining tactics and outcomes in both companies. In both, regular meetings were held at corporate headquarters with divisional or operating unit personnel directors, and sometimes with divisional managing directors, to discuss personnel policy matters. In particular, annual meetings at an appropriate time in the year before the bargaining round began were called to discuss bargaining strategy. Position papers were produced by corporate staff (leisure) forecasting changes in retail prices and other relevant measures and itemizing key factors in the company's financial and market position. In the foods company, the chief executive showed great interest in wage determination and tried to encourage the adoption of CBI norms for coming pay-rounds.

These annual pay-round meetings were essentially designed to help divisions reach a consensus on bargaining strategy and aims in line with corporate needs. In practice some hard bargaining took place over the mandate to be agreed. Changes in holiday entitlement, hours of work, and basic percentage pay rises were particularly contentious. Intra-organizational bargaining tactics became important since each divisional personnel director was anxious to preserve his bargaining freedom with the unions while protecting himself from dangerous or expensive precedents in negotiations and trying to keep his own position vague. In 'leisure' one divisional personnel director was 'unable' to produce his pay strategy paper for the coming year on the grounds that it had not been approved by the divisional board. The aim of corporate personnel executives at the meetings was to try to reach a consensus which agreed with the corporate budgetary needs and could be reported back to the board as an agreed maximum figure for payroll cost movements.

In 'leisure' emphasis on the need to keep within the budget approved by the main board meant that the divisional personnel directors were faced with 'cash limits' which restricted their bargaining freedom at a time when they felt they should have been given more discretion. This was resented and corporate personnel executives felt somewhat beleaguered, caught between divisions wanting bargaining freedom and the corporate finance needs of strict adherence to the budget. In the end divisions were told that they had to comply with the agreed mandate based on the cash limits. A pay strategy document was issued by the corporate personnel department to this effect after the mandate had been approved by the board.

Corporate headquarters in both companies expected to be kept informed on progress in negotiations with the trade unions, even, in one case, by daily telephone checks. They would provide tactical advice if required and occasionally agree to shifts in the mandate if no agreement could be reached and a revised offer was felt to be imperative. In 'foods' the corporate industrial relations manager would sit in on as many negotiations as possible but he had no power of veto. When he retired this practice was abandoned. On one occasion corporate staff in 'leisure' went to an operating unit which was experiencing difficulty in negotiations, but ensured that they stayed 'in the back room', unseen by the trade unions. Care was taken not to appear to weaken the authority of the local negotiators. Not only must divisions be seen to have to stand on their own feet but no opportunity must be given for the unions to bridge the divisional structure, it was argued. Interestingly, in each firm the unions did not appear to attempt to co-ordinate their bargaining, although on one occasion in 'foods' the unions walked out of negotiations in one division and refused to return until a better offer was made in another.

Why did corporate personnel staff try to co-ordinate and control pay bargaining? After all, there is little reason why operational management should not be left to get on with the negotiations, determining their own strategy in the light of product and labour-market conditions as though they were independent companies subject only (but a very powerful 'only') to the requirement that they keep to agreed budgets and targets. In practice five reasons were advanced for the need for mandate bargaining and consensus. First, wages and salaries constituted a major component of predictable cashflow and the corporate finance department felt it imperative to have decent forecasts for the coming year. Secondly, it appeared that the board always wanted to know what was going to happen in wage negotiations. As an exasperated corporate personnel executive in 'leisure' put it: 'I can't go to the board with no agreed figure from this (mandate bargaining) meeting and just say the boys will do a bloody good job as usual. They want a figure!'

Thirdly, there was public concern, as always, over wages and government was trying to 'talk wages down'. In 'foods' in particular the chief executive was conscious of his civic duty. Fourthly, it was hard to avoid the conclusion that co-ordination of pay bargaining provided a *raison d'être* for corporate personnel staff. It was one of the few areas where they could clearly intervene and provide some expertise and it seemed this had always been the custom. Finally, there was the fear of trade union co-ordination of bargaining strategy. Both corporate personnel staff and divisions recognized that an unduly generous (or mean) offer by management in one division could spill over into another area of the company, especially in terms of hours of work, holidays, and basic percentage pay rises. It was recognized that unions (and employees) are motivated in bargaining behaviour along the lines of relative wage justice, or 'a fair day's work for a fair day's pay', which is fundamentally at odds with managements' attempts to link pay to product and labour-market conditions. The pay parity strikes in the 1960s and 1970s were cited as evidence for this. Ironically, the weakness of the trade unions in the recession, and their surprising (to management at least) inability to co-ordinate decentralized bargaining in the two companies helped to weaken the authority of corporate personnel management. This was evident in the difficulty they had in influencing divisional management, and in terms of their authority at corporate level *vis-à-vis* the finance function. Not all M-form companies attempt such co-ordination and it is probable that the number who do so is declining, but control via the budget review process remains powerful. In effect central control and close monitoring remains, but it is more likely to be in the hands of management accountants and the finance function rather than in those of personnel (Marginson *et al.* 1993).

Our case-study findings revealed that what was in many cases labelled by companies as decentralized bargaining was in fact much more centralized and tightly controlled than first met the eye. For example, we found in two of our most avowedly decentralized bargaining structures that the link between decentralized bargaining and a profit centre logic was not as strong as assumed. Pay settlements tended to vary little between divisions and between plants in divisions. While there was an espoused philosophy from the centre of 'autonomy', this was reinterpreted by both divisions and plants as something considerably less than a freedom to do one's own thing.

This was reflected, in one of our cases, in a series of remarks made by different personnel managers at the three different levels of the corporation. We began by asking the central department what kind of control they exercized over bargaining activities throughout the corporation. The answer from the centre was 'no control'. We next asked personnel at the divisional level what kind of control was exerted over their bargaining. Their response was 'no control but some influence'. When we actually went down to plant level and asked the same question of plant bargainers the answer we got was 'full control'! Why this apparently contradictory series of comments? First of all, although plants were theoretically permitted to reach a pay settlement in line with their profitability, there was also at the same time considerable pressure to reach settlements based on guideline 'budgeted' figures, employed within the logic of the financial control system. There was, further, a considerable amount of 'informal' inter-plant discussion that took place over wage negotiation strategies which ensured a patterned settlement.

There was, in fact, much inter-division and inter-plant rivalry related to the size of the bargained wage increase and no one manager wanted to settle out of line with the corporate trend. Even though divisions or plants were theoretically free to settle where they wanted, there was a rather acute awareness and sense of what the 'appropriate' settlement ought to be. The organizational grapevine works as effectively here as in any other part of the organization. There was also considerable weight placed by divisional managers on 'words' from corporate office. Even the most vague words were picked up, queried, thought about, interpreted, and reinterpreted. For example, if the centre or board made some comment about the year being a 'belt-tightening' one this would signal to divisions that the next budget review would be austere, even if a particular division or subsidiary company was having a good year or had particular pay needs. In decentralized bargaining a key distinction is the level of actual negotiation and the degree of autonomy of the management bargainers. The blurring of the distinction between plant and corporate bargaining adds a further dimension to strategic choice around bargaining levels and exposes the choice as less clear cut and perhaps having more to do with appearance than substance. Kinnie (1987) has gone as far as to suggest that managers at the centre of the corporation promote the ideology of decentralization, behind which lies a reality of centralized control over major industrial relations decisions.

SUMMARY

Most large companies recognize trade unions. The 1985 company level industrial relations survey found that, of the companies covered, all of which employed more than 1,000 people, union recognition was widespread, with 89 per cent of the corporate respondents reporting that manual unions were recognized in all or some of their workplaces. For non-manuals it was three-quarters. The equivalent figure in 1992 was two-thirds. Thus most large firms engage in collective bargaining either directly or, less commonly, through the medium of multi-employer bargaining. And of course all companies, whether unionized or not, have to determine pay and pay levels. There is, in nearly all cases, an annual cycle to this and all large companies have to decide where pay determination is to be handled. Collective bargaining is at the heart of the relationship between unions and management, and pay is at the heart of the employment relationship. The decision on where this is to occur has fundamental consequences for the conduct of industrial relations and the design of personnel and human resource systems. Historically collective bargaining developed along multi-employer lines first at regional level and then nationally following the Whitley Committee reports at the end of the First World War (thus Whitley Committees were the prime negotiating bodies in much of the public service sector until recently). At that time, and for much of the post-Second World War era, industrial relations was not seen as a management function at all. It was external to the firm. It was the Donovan Report in 1967 which first authoritatively established that the management of industrial relations had been underdeveloped and called on companies to take industrial relations seriously.

The hallmark of the 'New Industrial Relations' of the 1980s was, at its simplest, that not only was industrial relations now seen as an internal company operation but that the conduct of industrial relations was made subservient to corporate and business strategy. Organizational inertia gave way to strategic thinking and the strategy–structure debate took on a real meaning as companies, public, private, and privatized, began to change their first- and second-order strategies. If industrial relations and collective bargaining had been fixed rocks around which companies had had to navigate in the past, in the 1980s they became movable and thus much less significant. This became more pronounced in the recession of the 1990s as the debate about bargaining levels took hold. This was, and still is, a debate almost exclusively within management circles. The unions have been remarkably silent, apart from a short-lived debate on co-ordinated bargaining (Industrial Relations Review and Report 1991), and have been forced in virtually every case to acquiesce. The one major exception was British Rail in the late 1980s, who failed to change bargaining levels. They were caught out by rising inflation, tight labour markets, and a macho management style which was totally inappropriate for the circumstances, but by 1992 progress had been made toward decentralization.

We have shown in this chapter how wider strategies and structures have combined to put pressure on the industrial relations systems and the extent to which bargaining remains co-ordinated within management. Most importantly we have shown how the prime force for bargaining decentralization was to be found within the M-form processes of diversification, divisionalization, and decentralization. These forces were developed outside the industrial relations arena but brought to bear on it. It is the clearest possible example of first- and second-order strategies influencing third-order strategies.

From our experience as researchers, teachers, and consultants in management we developed a tool to be used in helping to decide on bargaining levels, if only to bridge the academic/practitioner divide. The danger of such lists is that they come to be used mechanically, an approach which leaves out fundamental questions: why change? what is to be achieved? is it possible? will it work? what happens if things go wrong? We have often been surprised at how infrequently these questions are asked and at the wild assumptions made. The most important question is: what does the firm intend to achieve in bargaining decentralization? Is the move part of a considered change programme such that the style and pattern of industrial relations and of managing individual employees will be different in a material sense from those of the previous system? Will it be used to herald a new approach to management which might appropriately be called human resource management? This raises the question of management style; we turn to this in the final chapter.

7
Management Style in the Multi-Divisional Company

THE IMPORTANCE OF MANAGEMENT STYLE

The most distinctive and important feature of the strategic choice perspective is its focus on the capacity of organizational leaders to make choices at all three levels of strategy. This may seem unsurprising, indeed almost a truism. But within the field of human resource management specifically and the restructuring of organizations more generally it is a relatively new approach. In part, the explanation for this comes from the creation of the M-form company itself. Once a firm grows and develops into new markets and geographical areas the logic and efficiency of centralized functional forms breaks down. There is an information-processing problem matched with a lack of experience of those at the centre of the precise nature of the business, the market, and the technology in each area of activity. The M-form structure, with its emphasis on devolution and market orientation, allows each division, strategic business unit, and operating subsidiary to choose, to a greater or lesser extent, the most appropriate functional strategies for its own purposes. The centre is ill advised to seek to second guess what is best for a given SBU in terms of technology, marketing, production, and operational management. For this reason, more weight is placed on financial performance and less on administrative rules. Of course, as we have emphasized, the tightness or looseness of the financial targets set by the centre and its control over investment policy can deeply constrain and limit the type of choices that units can make. For example, 'short-runism' makes long-term investment in people much more difficult.

There are other reasons to emphasize the relative novelty of strategic choices. In markets with a large number of small players producing identical goods there is very little scope for choice in price or quality. The invisible hand of the market is the dominant force. Market maturity, higher costs of entry for potential new players as technology becomes more sophisticated, and the cost advantage of gaining market share all tend to reduce the number of players and each may seek to specialize in a particular niche. It is not very sensible, for example, to open a corner grocery shop alongside a major new supermarket with opening times of 60–80 hours per week and a weekly take of some £1,000,000. The dominance of large firms, especially in the UK, was noted in Chapter 2. Their power in the market allows them to make choices about their own organization, their borrowing requirements, markets, and product innovation. Of course, all choices are constrained. National and international governments seek to restrict monopoly power and market dominance by a single company, and have the capacity to change tax levels significantly. Growing international trade has opened markets to more competition, and the ebbs and flows of economic growth and the trade cycle directly impact on every company through the level of demand, interest rates, and consumer preferences. Yet with a range of businesses within the portfolio of a particular firm it is possible, to a degree, to offset losses in one with gains in another, to seek to enter a new market while exiting from a poor performing area, and to raise capital in ways often closed to the small entrepreneur.

We have emphasized throughout the distinction between strategic and operational management as one of the distinguishing features of the multi-divisional company. This is never as clear-cut as might be supposed, however. First, whatever the intention, it is often very difficult for corporate staff not to second-guess operating decisions, especially when budgets are not met and unit management promises not fulfilled. Senior executives in many cases have reached the top after years of experience in running operating units and divisions where they were expected to have a 'hands on', deep knowledge of markets and technology. It is hard for them not to seek to utilize the very skills that got them promotion to the top, and it is often the case that they use their experience in one division to second-guess the behaviour of another. This is as true of corporate personnel directors as it is of other senior staff. Indeed the injunction for firms to 'stick to the knitting' (i.e. to specialize in a particular area of product competence) and to build synergistic links between parts of the organization is, in part, to allow senior corporate executives to maximize the managerial competence gained in running business units. The greater the diversity of businesses in the portfolio the less this is possible, and for this reason some M-form companies have sought in recent years to divest peripheral activities.

Secondly, there is a clear distinction between corporate and business strategies, with the latter increasingly placed in the hands of divisional boards or subsidiary companies. Here the divisional board is concerned with the long-term development of its own business, which can include identification of firms to acquire, units to dispose of, and markets to exit from. It will also include decisions on the internal structure of the division (second-order strategies) in terms of operating subsidiaries, the level of profit centres, and the number and scale of establishments. Thus, the relationship between the divisions or major subsidiaries and the corporate board is likely to be complex and often tense. And, of course, each division and each operating unit will be required to develop its own functional, third-order strategies in marketing, production, and personnel, for example. The greater the diversity of businesses within the corporate portfolio the more likely it is that separate strategies will be required at lower levels. The more the corporation seeks to emphasize financial economies and financial management the less likely it is to seek to develop business strategies for all its units. The more the corporation creates synergies between parts of the enterprise and focuses upon a core set of activities, the more likely it is that the guiding force for strategy will emanate from the centre.

The third way in which the distinction between strategic and operational management is less clear than might be supposed is more germane to this chapter, and is found in the nebulous, yet critical, area of values, cultures, and styles. Reference was made earlier to the notion of 'institutional strategies' which seek to answer the question of what sort of firm is to be created and developed. Does the enterprise implicitly or explicitly make statements about its objectives to the community and stakeholders in terms of its underlying values? Some firms, for example some conglomerates, have reputations as 'asset strippers' while others are known for their concern for communities or the environment or for long-term investment in given markets. More specifically in relation to the management of people, what key values, if any, are articulated and how are these transmitted throughout the organization? This is the realm of management style.

Even if a particular M-form company decentralizes responsibility for personnel to lower levels of the enterprise, seeing personnel, as Hill and Pickering (1986) do, as an operational matter, it is likely that the values of the chief executive, mixed with the historical behaviour patterns of the firm and its traditions, will deeply influence the type of policies pursued and the style of human resource management. For example, strong anti-union beliefs held by the corporate directors are very likely to filter down to a given unit and affect the behaviour of unit managers in their dealings with the union officers and shop stewards. Similarly a tight rein on costs and doubts about the usefulness on training and development or the need for internal labour markets will limit the possibility of extensive investment in employees at the unit level: the preference will be to poach qualified employees from other firms. Even if the unit manager believes in the need for training and development it will be difficult to gain approval for such expenditure in a hostile budget review, especially where there are no reserve items required by corporate head office for training and development. If the only requirement in the budget review is the achievement of certain ratios, such as the rate of return on capital employed or sales revenue per person employed then training and development expenditure is seen to be discretionary and is likely to be forced out in order to meet these bottom-line targets. Thus, although in theory employee relations and responsibility for human resource management may be formally vested in the hands of unit managers, in practice the corporate headquarters, with or without a personnel director, are deeply influential if only in giving clues to what is expected and what is disallowed.

At its simplest, corporate culture is 'the way we do things here', as revealed in stories of past failures and successes, the myths and meanings attached to events, rituals, and symbolic actions. Powerful signals are sent from corporate offices, often in covert form, of what is expected, rewarded, sanctioned, permitted, and prohibited. The tendency for middle management to treat corporate executives as gods often seems extraordinary to those unfamiliar with corporate life, and comes as a shock to many students entering the world of employment for the first time. The messages passed down from Mount Olympus by what Winkler called 'the ghost at the bargaining table' (1974) inevitably set the parameters for management style and behaviour towards employees even if they are unwritten and often, indeed, unspoken.

In one of our companies (staple products) for example, we discovered what was called 'management by raised eyebrow'. One company director put it this way:

> It's impossible to have a uniform style across the entire group. But, still, certain aspects of corporate style tend to drift down to the various businesses. These could be defined as 'practical', 'commercial', and 'tough minded'. Through all of this autonomy they do get a drift of the style. The understanding of our style is a result of a series of questions asked by divisional head office [of the corporate office]. It is through a series of small and insignificant questions that they learn to behave at the local level and put themselves in sync. with the wider corporate body.

We were given a vivid example of this 'style' in practice. The corporate personnel director intervened directly in a staffing problem resulting from the amalgamation of two divisions. A decision had to be made as to which of the personnel managers of the two divisions would be kept on as the head of personnel of the amalgamated divisions. In this case the central personnel director had a choice between two individuals: one typified the hard-nosed, economy-driven approach of the parent, but had limited personnel experience, the other represented a more caring, and more 'professional', approach to personnel. The former was an ex-line manager while the other was a personnel professional in the true sense of the word and held high standing in the Institute of Personnel Management. In this case the personnel director at corporate level, for purely strategic reasons, decided against the personnel professional and hired instead the 'hard-nosed' ex-line manager. He justified his decision on these grounds:

> He was a pure personnel professional [talking of the person who did not get the job]. He was driven by personnel considerations only. In our company we try for a more balanced approach to personnel in which personnel is tied to the business. Personnel people must be business people first. We don't want substitute trade union officials as personnel

managers; a good personnel person does not equate to IPM personnel professionalism.

It can clearly be imagined how the message of this appointment swept through the whole of the division and beyond to all of the personnel managers in all of the units and the sites in the company. Personnel professionalism was not highly valued; cost minimization and hard-nosed business decisions clearly were. This also points to one of the prime ways in which style is articulated and inculcated. The critical and key role of the corporate personnel department, as described in Chapter 5, is the appointment of senior managers at divisional and operating subsidiary level. This is a visible power and the type of appointment and the distinction between those that gain and those that lose sets an unmistakable style for management within the whole of the corporation. If you want to be successful you need to modify your own behaviour to fit the corporate style.

This became clear to us in a different sense in our foods company. Here a divisional managing director had recently been appointed from outside the company. There appeared to be no particular reason why he should have left his old firm to join the foods corporation. He explained this in terms of getting to know, in his previous business dealings, some of the senior managers in the foods corporation and 'liking their style': 'I thought this was a humane organization with an emphasis on efficiency and growth. It seemed a good place to work in. This was a company which had made greater efforts than most, largely because of the personal values and energy of the chief executive, to develop a distinctive management style.'

The importance of style is now more evident than it was because of the rapid growth in the need to make strategic choices in human resource management. This has come about for two reasons. First, in many countries there has been a progressive decline in national systems of industrial relations and a growth in firm-specific employment policies. This is especially evident in the United Kingdom (Purcell 1993). For much of the earlier part of this century industrial relations was considered a completely external factor to the firm. Pay and conditions were set by employers' associations and confederations of unions within the forum of multi-employer bargaining and jobs were simply graded, with individuals having generic skills. For as long as stable conditions and government support for national bargaining arrangements existed there was little need to think about making choices in employee relations within specific firms (Gospel 1992). This was, and remains, especially the case where legislation on pay spreads the collective bargaining agreement throughout the whole of a particular industry, as it does for example, in Germany.

This reliance on external institutions and procedures of industrial relations has now largely disappeared in the UK, as we noted in Chapter 6. Very similar trends are occurring in the USA, Australia, and New Zealand, and, to a lesser extent but clearly visible, in many countries within Europe such as Sweden, Norway, Denmark, Germany, the Netherlands, Italy and France (Ferner and Hyman 1992). Everywhere there is a decline in the comprehensiveness of national agreements for industrial relations and, as we showed in Chapter 6, in the UK there has been a general collapse in such arrangements. There is also greater diversity emerging between firms within the same country, and it is hard now to talk of the industrial relations system of a particular country. This is especially the case in those firms facing international competition, or where a national government is forcing through a programme of radical change in the public sector, as is the case in the UK. If national systems are breaking down, firms are compelled to make choices, and to internalize and take responsibility for the management of employee relations.

The opportunity to make choices in employee relations is also related to the decline in trade union power and the reduction in trade union membership from its high point in the 1970s. In the UK at least, this is in turn related to legislative change, and more especially to the rapidly changing composition of the workforce in a context of long-term unemployment. The dominant picture of trade union members as full-time manual men in the manufacturing and extraction industries is changing in the face of shifts in the sectoral composition of the workforce into one of part-time and temporary employees, women workers, and a substantial growth in the white-collar and service sectors. It is here that unions have faced enormous difficulty in retaining members and in recruiting new ones. The outcome is that union membership within the corporation is likely to have declined and closed shops, which were relatively widespread in the mid-1970s, have now all but disappeared in practice, as well as being outlawed by legislation (Millward *et al.* 1992). Table 7.1 shows the shift in trade union membership internationally in the 1980s. The effect of this decline in both membership and power has been to open up choices to companies on how they should approach the question of union recognition and membership encouragement in both new and existing sites. Within the UK they will face further choices on the extent to which they wish to support the automatic deduction of union dues from the pay packet (the checkoff) as the 1993 Trade Union Reform and Employment Rights Act comes into force.

TABLE 7.1. *Trade union density in selected countries 1979–88 (rank order in 1988)*

	1979(%)[a]	1988(%)[b]	% change
Sweden	89	85	–4
Denmark	86	73	–13
Finland	84	71	–13
Norway	60	55	–5
Belgium	77	53	–24
Ireland	49	52	+3
Austria	59	46	–13
Australia	58	42	–16
UK	58	41	–17
Italy	51	40	–9
Canada	36	35	–1
Germany	42	34	–8
Japan	32	27	–5
Switzerland	36	26	–10
Netherlands	43	25	–18
United States	25	17	–8
France	28	12	–16

Sources:
[a] Kelly 1989.
[b] OECD Employment Outlook, July 1991.

An example of the type of choices being made comes from our privatized corporation which, like all other privatized companies, had high-density union membership at all levels of the firm. The firm took the view, once it had been privatized, that union membership amongst middle managers inhibited responsiveness to change and implied or imposed a clash of loyalties between the union and senior management. The evidence for this was pretty thin but such beliefs often have the quality of a 'factoid': if you believe it, it is real to you. Individual contracts were offered to all managers and many received a company car when they signed. Most, not surprisingly, did; only a few old stalwarts chose to remain on the existing collective contract. The union still exists and has a surprisingly high membership amongst managers, but it is no longer recognized for collective bargaining purposes. Its prime role is now to act as a pressure group on behalf of its members, issuing surveys and questionnaires, managing the media, and representing individual managers who have a grievance or a dispute.

Similar choices confront an enterprise when it embarks on a greenfield site investment. Here there is a clear choice to make on whether to encourage a trade union to recruit members or to operate as a non-union plant. If the decision is to recognize a union, a second choice then has to be made on which union to choose. This is the so called 'beauty contest' where unions compete with each other, not for members as in the old days, but to gain 'permission' from the employer to recruit, since only one union will be chosen and the power to choose is vested exclusively in the hands of the employer. Thereafter there is a further choice in the type of relationship the firm will seek to pursue with the recognized trade union: whether to emphasize partnership and co-operation with extensive exchange of information, or to have a hands-off, more adversarial approach limiting the interaction between management and the union as far as possible to annual pay negotiations. Within existing operations, where union membership falls, a time eventually comes when a firm will choose whether to continue to recognize the union, the nature and quality of the relationship, and the scope of interaction and discussions.

In all of this the view of the corporate office will be of prime importance. Every firm has a strategic choice to make in managing employee relations in ways that seemed unimaginable thirty years ago. The 1992 company-level survey shows that 60 per cent of the companies issued instructions on their 'style' in managing employees, 20 per cent gave advice, 7 per cent provided guidelines, and 13 per cent left it up to operating units or divisions (Marginson *et al.* 1993).

There is a second sense, and perhaps a more profound one, in which choices that deeply affect management style now exist, or are recognized to exist, in the area of employee relations and especially human resource management. The certainties of the past about the best way, indeed for many managers the only way, to manage employees have disappeared. The work of F. W. Taylor and the scientific management movement had one of the most profound impacts in the twentieth century on the way managers think about the world of work and employment, with its notions of management supremacy, the fragmentation of tasks and the command-and-control type of supervision. The challenge to the deeply embedded assumptions about the best ways to manage is perhaps best put by quoting a statement from a senior Japanese business leader made in the mid-1980s. It is of particular interest to note that this statement is now widely used in some company training programmes for managers in the West as a means of learning about and responding to the 'challenge of Japan'. (It is, for example, well known in the Rover Motor Company as it struggles to survive now that Nissan, Toyota, and Honda have all opened manufacturing, car-assembly and engine plants in the UK.)

THE CHALLENGE FROM JAPAN

We are going to win and the industrial West is going to lose: there is nothing much you can do about it, because the reasons for your failure are within yourselves.

Your firms are built on the Taylor model; even worse, so are your heads. With your bosses doing the thinking, while the workers wield the screwdrivers, you are convinced deep down that this is the right way to run a business. For you, the essence of management is getting the ideas out of the heads of the bosses into the hands of labour. We are beyond the Taylor model: business, we know, is now so complex and difficult, the survival of firms so hazardous in an environment increasingly unpredictable, competitive, and fraught with danger, that their continued existence depends on the day-to-day mobilization of every ounce of intelligence.

For us, the core of management is the art of mobilizing and pulling together the intellectual resources of all employees in the service of the firm. Because we have measured better than you the scope of the new technological and economic challenges, we know that the intelligence of a handful of technocrats, however brilliant and smart they may be, is no longer enough for a real chance of success. Only by drawing on the combined brain power of all its employees can a firm face up to the turbulence and constraints of today's environment.

This is why our large companies give their employees three to four times more training than yours, this is why they foster within the firm such intensive exchange and communication; this is why they seek constantly everybody's suggestions, why they demand from the education system increasing numbers of graduates as well as bright and well-educated generalists, because these people are the lifeblood of industry.

Konosuke Matsushita, 1985

This is an idealized statement of Japanese employment practices, even within the large-corporation sector of Japan whence it originates. It also grossly stereotypes western approaches to people at work. But there is enough truth in it to be hard-hitting, and the fact that is has now achieved the status of one of those statements which is passed from hand to hand is indicative of the sense of uncertainty that surrounds established nostrums of western employment relations. This explains, in part, the quest for new thinking and big ideas that has taken hold and is visible in the rapid diffusion of the language (but not the practice) of human resource management, total quality management, and even the direct use of certain Japanese words such as keizen (continuous improvement) (Oliver and Wilkinson 1992). All of this points to a reappraisal of the contribution that employees can, and should be able, to make to company performance while simultaneously, it is argued, achieving personal growth and satisfaction (expressed most vividly in the use of the word 'empowerment'). This is closely analogous to the ideals of human resource management described by Guest (1987) which were explored in an earlier chapter.

The problem with such descriptions, and with such terms as human resource management and total quality management, is that they are taken out of context and described in ways which appear to mean that they can be applied to every work situation. Like the firms described in *In Search of Excellence*, human resource management becomes attractive because it presents 'a coherent, positive and optimistic philosophy about management . . . built around the possibility of achieving personal growth in an integrated, human organisation' (Guest 1992: 17). The very attraction of human resource management is that it appears to fit the values and beliefs of managers and professional employees. We need some sort of map to help guide choices of the most appropriate style to suit the circumstances of the whole firm and its individual units. This is the function of the management style matrix.

THE MANAGEMENT STYLE MATRIX

How, then, are we to judge what style is most appropriate for a given unit, and what overall principles guide corporate philosophies about the management of employees?

Many writers have sought to distinguish between different types or patterns in industrial relations in terms of a conflict– cooperation spectrum. An important analytical distinction must be drawn, however, between classifications of the outcome of the interaction of management and employees, and the attitudes, beliefs, or frames of reference of the parties, most notably management, in determining the style that they wish to pursue. Clearly, frames of reference are influenced in varying degrees by the historical experience of managing employees in the firm and by wider social and political values. If management style exists at all beyond historically determined reactive gestures, we have to allow for an element of choice which might be more or less constrained. Thus the study of management style is not primarily an analysis of outcomes but of the philosophies and policies which influence action. Management style is different from management attitudes. We cannot make the assumption that attitudes necessarily translate into action or behaviour.

A good example here is the work of Poole (1981) and his colleagues who posed generalized statements in their attitude survey of managers about, among other things, the role and power of trade unions. We know from this that most managers are—or were at the time of the survey—distrustful of trade unions and the role they played, but we do not know what, if any, policies or preferences they had on how unionized employees should be managed in their own firms. Similarly the existence of certain structures, for example of negotiation and consultation, cannot be taken, as MacInnes (1985) has shown, to indicate the widespread adoption of a pluralist frame of reference. Our concern is only in part with how management chooses to respond to the existence of structures of collective bargaining, whether it opposes or supports them. If a preference exists (and there is no reason why it should) it must be, to a degree at least, translated into policy or determined behaviour for us to be able to say that the company has a distinctive management style. It is not necessary for such policies *always* to lead to the desired outcome but it is likely to exclude the adoption of certain types of behaviour. Indeed, given the variety of circumstances within the corporation, we would expect to find differences in practice, but guided by the same underlying style principles coming from the corporate office. We would also expect there to be a frequent gap between aspiration and outcome, if only because style statements tend to be normative and idealistic.

Style implies the existence of a distinctive set of guiding principles, written or otherwise, which set parameters to and signposts for management action regarding the way employees are treated and how particular events are handled. Management style is therefore akin to business policy and its strategic derivatives. Indeed management style is one of those aspects of wider business policy which 'state in broad terms both what may and may not be done . . . and . . . are more often made as a result of moral, political, aesthetic or personal considerations than as a result of logical or scientific analysis, and are usually made by the owners or the directors of a company rather than by executives at the lower or middle levels' (Argenti 1976: 63). Not all firms have a defined business policy in the sense of a mission statement or guiding purpose, and many of those which do say nothing about the management of employees, seeing such matters as an operational responsibility of middle management. Pragmatic, reactive responses to labour problems cannot be classified as management style. Indeed, in the extensive literature on business policy and corporate strategy it is extremely rare to find any reference to employees, personnel management, or human resource strategy. Thus the study of management style in employee or labour relations is not to be confused with analysis of management practices in each and every firm.

Fox's use of unitary and pluralist frames of reference had a major influence in initiating the debate on the approach taken by management in industrial relations (Fox 1966, 1974), but the concepts are narrower than those implicitly suggested by management

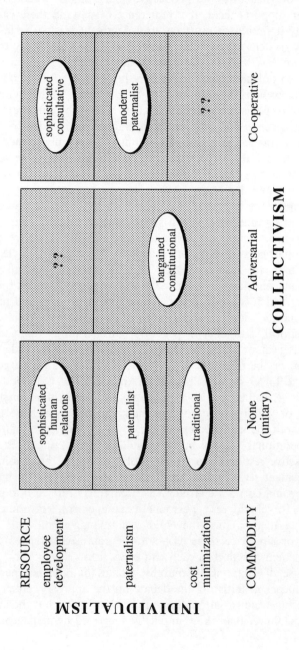

Fig. 7.1. Thew Management Style Matrix

style and have a number of limitations. First, wide variations can be found within the unitary frame, based around a single source of (managerial) authority, between those firms which are essentially exploitative of labour and those which emphasize the value of loyalty and commitment. Similarly, pluralism, where differing goals are reconciled by negotiation, needs to be subdivided into those managers maintaining highly structured, often antagonistic, formal relations with employees through their trade unions and those which emphasize dialogue, understanding, and co-operation with organized labour.

Secondly, and most importantly, the very terms unitarist and pluralist do not seem useful as a means of articulating the complexity of management styles since they are, by definition, mutually exclusive. The way the terms are used means that a pluralist firm will recognize unions while a unitarist one will not. Of course, it is quite possible that a firm where unions are recognized will have managers who hold unitarist views (Poole *et al.* 1981). Thirdly, it is often unclear, in the way the terms have been used subsequently, whether they relate to management's beliefs and policies toward trade unions or, in addition, cover direct relations with employees. In practice, as well as in theory, firms may have broad policies or guiding principles regarding both the attention given to employees and their individual and group needs, and that given to trade unions and other types of collective labour organizations. Such policies are not necessarily mutually exclusive. Thus two dimensions of management style need to be identified which we might term individualism and collectivism (see Figure 7.1).

Individualism

The component characteristics of the style matrix need to be explained. The vertical axis relates to individualism. This raises the question of the extent to which the enterprise views its employees either as individuals with needs, aspirations, competencies and particular skills of their own, or treats them as homogeneous blocks of people, with personnel and payment policies unable to distinguish between individuals and individual performance. Thus at the top of the individualism scale, which might be described as high individualism, the expectation is that the firm thinks of its employees as a 'resource' reflected in the, often banal, statement 'our employees are our most important resource'. Another way of describing high individualism policies is to use the term 'internal labour markets'. In this system the corporation has narrow ports of entry, for example taking in just graduates and school-leavers and seeks to develop individual employees through internal promotion patterns and by giving access to training and development.

The contrast, at the bottom of the individualism scale, is the view that employees are essentially a commodity to be bought from the external labour market and, where economic need dictates, returned to the unemployment pool. Here we would expect to find that the firm recruits individuals with a basic set of skills as and when it requires. It is unlikely to embark on extensive training and development and is more likely to opt for low-skilled employees. It would be here that we would also expect to find emphasis to be placed on numerical flexibility and the extensive use of temporary, part-time, and seasonal workers typical of hotels. This is described as a 'cost minimization strategy' where the emphasis is placed on short-term employee costs, a tight control over pay and employment levels, and a preference to avoid training and development.

In the centre, the category 'paternalism' refers to companies with a tradition of welfare-based personnel policies which seek to emphasize loyalty on the part of employees, often by means of reasonably generous fringe benefits and pay levels. These firms are likely to use such words to describe their style as enlightened, benevolent, charitable, welfare, caring, humane, paternal, family, happy, and asset. These words were taken from the first company-level industrial relations survey (Marginson *et al.* 1988) where we asked firms to describe their overall style statement. This contrasts with words such as firmness, discipline, costs, effective, productivity, manpower, strong management, and performance which were more akin to the cost minimization style. Paternalists have little expectation that employees will be promoted or developed beyond their original or existing jobs, and what training is provided is directly concerned to improve their performance in their existing task. Two quotes illustrate the general theme. One corporate-level manager described his company style as: 'Benevolent paternalism. There is a caring feeling. We should take care of our staff.' Another manager said that his company had 'Generated a caring, paternalistic attitude. It listens to employees. It is tolerant of the employees within the constraints of successful business.' The *Oxford Dictionary* definition of paternalism is revealing: 'limiting the freedom of the subject by well meant regulations'. Companies with a paternalist approach do not appear to place emphasis on employee development and career progression and other attributes of individualism, nor are they dismissive of a sense of social responsibility toward their staff. Notions of caring, humanity, and welfare are emphasized as a means of legitimizing managerial authority and the subordinate position of lower-level employees, who are given few, if any, expectations of a possibility of changing their work roles and their place in the hierarchy. The corollary of paternalism is thus the deferential worker.

'Employee development' at the top of the individualism scale describes the style of those firms which seek to develop and encourage each employee's capacity and role at work. A typical quote from the 1985 survey is:

Our company's paramount concerns are recruitment, development, pay and safety of employees. We aim to recruit above-average employees who expect to make an above average contribution. We invest time in development and we very much promote from within the company.

Another respondent at the corporate level said:

We are quite unusual, operating as the leaders of teams. We believe very much in individuals; we trust everybody. We work on the premiss that every person needs to have the opportunity of self management. Hence, the manager's role is very much as leader of the group.

It is not surprising to find that the 'soft' definition of human resource management is very much akin to the style of employee development with this emphasis on teams and teamwork, competencies, appraisal and reward, empowerment, training, and development.

Marchington and Parker (1990) tried to test the original style matrix (Purcell 1987) and found it wanting in a number of respects. They suggest that the individualism scale could be simplified by testing the extent to which 'management' adopt an 'investment orientation' towards labour (Marchington and Parker 1990: 234). This is a useful modification, since it is clearly the case that at the resource end of the scale the firm is more likely to invest in human capital formation while simultaneously seeking to control labour costs. However, cost control is more likely to be based on unit labour cost control in the sense of policies to maximize productivity or output per person employed—sometimes loosely described as a high wage–high productivity strategy. At the commodity, cost minimization end the approval is more focused on the wage bill than on productivity per person employed: a low wage–low productivity strategy. The former implies investment in both capital and people; the latter requires the exploitation of existing resources.

Neither Marchington and Parker, nor more recently Storey (1992), like the term 'paternalism'. They note its static quality and the way it could be applied to firms which are in the process of change (Marchington and Parker 1990: 234). Historically, however, there is clear evidence that some firms have developed paternalist styles towards employees. The fact that, as we shall see, paternalism is being driven out as firms seek to respond to turbulent product market conditions and financial constraints does not invalidate the categorization. Storey complains that the model is essentially static. This is true if it is only used to describe a given pattern at a given point in time. But used as a map for the development of strategic choices in employee and human resource management, and for tracing developments over time it can provide a method of capturing dynamic movements and plotting strategic intentions. This is explored later in the chapter.

Collectivism

Storey and Bacon (1993) also assert that the terms individualism and collectivism are used in the matrix in a narrow industrial relations context. They argue that 'collectivism' can be used in a Japanese sense, to describe group unity, seen most vividly in team-work. There is much in what they say. Collectivism can be used to describe certain forms of work organization based around teams. The problem is one of language and definition. The term collectivism here is used in its political and industrial relations sense, closely associated with pluralism. Here collectivism refers to the process of forming voluntary associations, the collective, first to aggregate and then articulate individual interests through a collective whole, forming secondary institutions that act as interest- or pressure-groups and seek a role in decision-taking with more powerful decision-takers. The critical question confronting those with power, in this case management, is whether to oppose, condone, or encourage the emergence of the collective labour organization, and to ask what sort of relationship should be developed with it. Pluralism in this sense is about power sources and power-sharing.

That this use of the term collectivism is preferable is even clearer when we recognize that team-work, as a different type of collectivism seen in a work organization, frequently requires the prior identification of each person's capacity, competence, and willingness to be part of a team. In practice, management seeking resource-based employment policies where team-work is seen as crucial, spend time and money selecting each individual, providing training, and appraising performance such that only 'good' team-workers are encouraged. They thus exhibit 'high' individualism in their employment policy.

The horizontal axis of the matrix is used to describe the company's approach to aspects of collectivism. The term collectivism is used here to describe company policies which relate to the recognition that employees have the right to form themselves into independent or quasi-independent organizations and to elect representatives to work on their behalf. In most cases this will clearly be the recognition of the legitimate right of trade unions to organize, represent, and negotiate. But it can also mean the active encouragement of staff associations, joint consultative committees, and, within many parts of continental Europe, the existence of works councils. Here, employees elect, by right, some of their number to work in concert with management, representing the views and interests of employees alongside the trade union. The critical question on the horizontal axis is the extent to which the firm gives credence, if at all, to the role of collective labour organizations. On the left of the scale the terms 'none' or 'unitary' refer to those firms which deliberately choose to avoid any form of collective labour organization and certainly seek to resist either coercively or by competition the possibility that unions will gain a foothold in the company.

In the central category marked 'adversarial' are those firms which have long-established relationships with trade unions but where the relationship is essentially of an adversarial nature. This implies that the prime activity is bargaining, with either party seeking to restrict information flows to the other and a tendency to exaggerate positions in order to reach a compromise agreement somewhere in the middle. It is important to recognize that the historical roots of adversarialism go far beyond the sphere of industrial relations. The legal systems in the UK and USA are predicated on an adversarial relationship between prosecution and defence. Party politics in the UK (and Australia) are similarly organized along adversarial lines, as is seen most vividly in Prime Minister's Question Time in the House of Commons, where there is an exaggeration of differences and certainly little joining of minds.

The contrast, a co-operative relationship, shown on the right of the matrix, is where words such as partnership might frequently be used to describe the nature of the relationship between the collective labour organization and management. This is likely to be seen most clearly in the extensive exchange of information, the use of joint working parties to explore problems and to handle difficulties of job grading, and in extensive efforts by either party to support the other. One example of this may well be that of an electronics company in the UK which sent a delegation of union shop stewards and managers to Japan to visit the major Japanese competitors with a view to improving the production and operational management system in the UK base. This is a classic case of a joint problem-solving co-operative style of employee relations.

Linking Individualism and Collectivism

The key proposition within the management style matrix is that organizations deliberately choosing a particular style have to operate on both axes of the matrix simultaneously. That is to say they have to decide on the appropriate level of individualism in association with the required type of collectivist approach to organized labour.

For example, in the foods company union recognition was granted in the late 1960s, with two unions recognized, one or the other in each of the six plants. When single-employer bargaining developed, the machinery of joint negotiations was located at the corporate level under the responsibility of a director of industrial relations. His job was to 'manage' the full-time officers both formally and informally, and the chief executive would from time to time invite union leaders to join him over dinner, or even in the corporate jet! Personnel matters, meaning the recruitment and deployment of labour, were the responsibility of the personnel director. Thus functional differentiation reinforced the separation of policy. In terms of the matrix, the personnel director looked after the vertical axis (individualism) and the industrial relations director the horizontal axis (collectivism). Of course, they claimed to co-ordinate action, but in effect the role of the industrial relations director was to keep the unions away from personnel and operational matters such as aspects of work relations (the jobs and tasks people do) and employment relations (the number of people employed, the way they are recruited, and the type of contractual status they have). To put it another way, the industrial relations director managed the external boundary with trade unions seen as an external environmental force, while the personnel director managed the internal labour supply.

This separation gradually broke down in the 1970s as shop stewards at the plant level began to question whether the full-time officers really understood their needs. In line with unions in many other companies, they began to demand a place at the bargaining table. At the same time, partly in response to government incomes policies which emphasized productivity improvements, and partly because the pace of change in technology grew, managers at plant level felt the need to build relationships with the shop stewards, give them more facilities and training, develop joint consultative machinery, and build plant-specific employee relations policies.

During this period the foods company diversified, divisionalized, and decentralized in a way typical of M-form companies. And, as described and analysed in earlier chapters, it broke up the corporate bargaining machinery and placed responsibility for industrial relations in the hands of profit-responsible plant directors. Local negotiation and consultation meetings are now taken up with the introduction of technical change, access to training, multi-skilling, reduction of numbers employed, the type of jobs people are asked to do, the design and operation of payment systems, and the use of part-time and subcontracted labour alongside full-time employees. Thus industrial relations is internalized at the workplace level and is combined with aspects of personnel management, under the guise of employee relations and human resource management (the terms now used). The prime pressure for productivity improvements and changes comes from production managers, and from the production and operational management systems.

The greater the decentralization of industrial relations, the greater the emphasis placed on workplace representatives, and the greater the pace of change in work relations and employment relations, the more likely it is that strategic choices in management style will combine individualism and collectivism. That is to say that action needs to be taken simultaneously on aspects of individualism and collectivism. It is necessary to combine decisions on work relations, employment relations and industrial relations, and link these to business strategies and corporate policies. Management style requires policy integration in a way that was unusual a decade or more ago. This requirement for integration is visible in many countries, as national systems of industrial relations break down and emphasis is placed on what Dore (1989) has called 'organizational-specific employment patterns'. Trade unions are parochialized while internal labour market structures are given more emphasis in those companies developing long-term employment patterns based on employee development.

Combining the two axes of the matrix, we can look at the characteristics of each of the boxes. Each is given a short descriptive title. The empty boxes with ?? in them describe conditions which are inherently unstable and are unlikely to last for long. For example, the bottom right-hand side of the matrix depicts a situation where relationships with the union, shop stewards, or members of the works council are friendly, co-operative, and based on principles of partnership. However, at the same time the firm is essentially exploitative of its labour, basing its policies around the achievement of cost minimization. This implies holding down pay to the lowest possible level, avoiding promises of job security, and being extremely reluctant to engage in training or make welfare payments. Such a firm is likely to emphasize numerical flexibility through deskilling and employment of

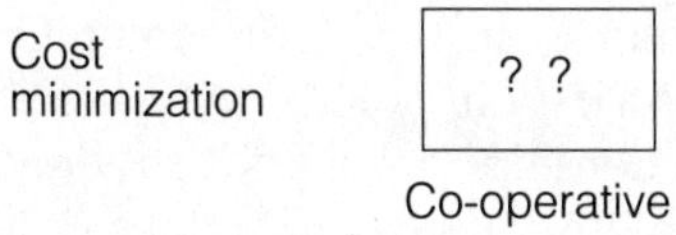

Management Style in the M-form Company

```
            Cost          ┌─────┐
         minimization     │ ? ? │
                          └─────┘
                         Co-operative
FIG. 7.2.  Unstable Style I
```

low-skilled employees. How, in these circumstances can a union or works council be expected to engage in highly co-operative behaviour? If it does so it is unlikely to be representing the interests of its members, and members will either leave, since nothing is achieved by staying, or may eventually revolt by electing new works councillors. Managers who have expectations that they can build co-operative relations with union officers, especially shop stewards, in these conditions are likely to be deluding themselves. This is an obvious danger in a country like South Africa where firms seek 'high trust' relations (Purcell 1981) with union officers representing black workers whose living and working conditions are usually extremely poor, and who have had virtually no rights at work (or in society) in the past. And for unions to co-operate fully in these circumstances is extremely dangerous for their own credibility and long-term survival.

The other box which represents highly unstable conditions is where relationships with the union are marked by their adversarial nature yet the firm is pursuing policies towards employees based on principles of employee development. Discussion of this segment is best left until after we have explored the characteristics of the other parts of the matrix and debated the way in which changes in style can be managed. It is merely necessary at this point to note that the way the firm seeks to manage its employees has a strong influence on the quality of the relationship with the trade union or works council. Highly adversarial policies towards trade unions will make it significantly more difficult to achieve progress towards high individualism, if only because it is likely that an impasse will be met in negotiations. The unions are likely to view policies of direct communication with employees (their members) and the use of appraisal and performance-related pay, for example, as essentially anti-union initiatives. They will feel bound to try to stop such policies since they challenge their historically defined role and existence. The failure of quality circles in the Ford Motor Company in Britain in the late 1980s is a case in point.

Adversarial relationships are marked by low trust where there is doubt about the other party's intentions (a tendency to believe the worst) and where, in Luhman's terms, people rely 'more and more on less and less' (1979). They require proof and do not take things on trust, while simultaneously being unable to rely on the sources of proof offered, following the famous phrase 'they would say that, wouldn't they?'. This is especially a problem in large employment units (Allen and Stephenson 1985). All this means that it is especially hard to move from adversarial positions which are culturally embedded without carefully considered integrated change programmes.

The Components of Management Style

Before we can explore this we need to elaborate further the nature of the boxes. It is important to bear in mind that the descriptions of the characteristics in each box are ideal–typical abstractions, which extract the essence of the particular style in order to bring it into clear focus and capture its key characteristics. Reality, as always, is more complex and less clearly defined. In policy terms the boxes do not list all of the relevant personnel and HRM policies nor is it the case that a firm with a style shown in a particular segment will have all the behaviour patterns listed. The purpose is to provide a map to indicate the various types of management style and to try to relate these to particular types of firms and particular labour and product market conditions.

Sophisticated Human Relations

A good starting-point is the box labelled 'sophisticated human relations'. This matches the 'soft' or empowerment commitment definition of HRM described by Guest (1987) (Figure 7.3). Here most employees, excluding those on short-term contracts and temporary or subcontracted labour, will be viewed as the company's most valuable resource with an expectation that most will choose to work for long periods with the firm. As a result great care is taken with recruitment and selection; this is evidenced in the use of competency profiles and psychological testing. Pay

```
employee              | sophisticated
development           | human
                      | relations
```

no collectivism
(unitary)

Fig. 7.3. Sophisticated Human Relations Style

levels are likely to be set above the regional going rate and many firms with this style have deliberate policies, based on regular regional salary surveys, to pay in the upper quartile of the relevant pay distribution. By definition, then, only a few companies meet the strict requirements of this style statement. Internal labour market structures are likely to be emphasized, with promotion ladders and outlines of career systems in place. Great care is taken to harness employees' views through the use of periodic attitude surveys, exit interviews, and open communication systems. Emphasis is placed on flexible reward structures, employee appraisal systems, and development and training plans.

The key issue for management is the assessment of *each employee's* competence, capability, and conduct in order to plan future activities. There is a clear need to recognize the contribution of *each* employee, including the capacity to work in teams. The organization of work and tasks is likely to include teamwork and total quality management systems, and words like 'empowerment' will be commonplace. There will be formal systems for employee grievance handling, open-door policies, disciplinary rules, and consultative procedures, allowing individual employees to express their views. Extensive communication channels, from induction films to company newspapers and videos, open days to company social events will be used.

The aim is to inculcate employee loyalty, commitment, and dependence and thereby to maximize productivity and responsiveness to change. High expectations are placed on employees and there is little spare or slack time (or 'porosity' as Elger (1990) calls it). Work is thus intensive and employees are expected to work beyond their contract in terms of compulsory overtime, going beyond the job description and helping others in their tasks. A by-product is that these companies seek to make it unnecessary or unattractive for staff to unionize, but if they do, union recognition is unlikely to be granted and action may be taken to reduce the threat.

Where are we likely to find this style of management? We can answer this in three ways. In terms of a generic style which seeks to capture the essence of a company's policy towards employees, in terms of the type of company, classified according to the characteristics of the dominant type of labour force, and in terms of the occupational classifications where this style is likely to be adopted. This may appear to confuse the issue but companies may have different styles applicable to different types of employees. For example, there is often a considerable difference between the styles adopted for managers as employees and those adopted for manual and lower-skilled clerical workers. In some cases there may be markedly different styles for semi-skilled workers, skilled people with a craft training, technicians, and supervisors, or even between departments based on the leadership style of the manager. If this is the case we would be bound to argue that there is no deep-seated management style and that, in effect, the firm is essentially opportunistic. Style, to be meaningful, must describe the dominant and all-embracing approach a company takes to employment relations.

With this caveat in mind, the sophisticated human relations approach (or what we might now call soft human resource management) in non-union firms has most often been found in American-owned electronic and information firms such as IBM and Hewlett Packard. Other types of firms where we expect to find this style are high technology firms where the capital: labour ratio is markedly in favour of capital intensity. These firms, with their relatively few employees and their high productivity or sales value per person employed can often afford to invest in people. The nature of the diagnostic work that employees undertake, where the firm relies on their judgement and problem-solving skills (as opposed to simple motor skills) means that such firms need to employ high-quality individuals and to invest in their training and development. It is also very likely that the skills they need for their jobs are highly specific to the firm and cannot be directly recruited from the external labour market. It takes time and training for them to learn what is required in their jobs such that the firm must invest in them. This would also be true of professional service firms such as management consultancies, accountants, merchant bankers, and legal partnerships.

This style is also much favoured in green field sites where the attempt is made, via technical sophistication and a new form of employee relations, to break away from the rigidities of traditional manufacturing establishments and, by being non-union, to give management significantly more freedom in managing the plant. Finally, in many large, well-established firms this is an accurate description of the style the company adopts in managing managers and the essence of their management development programmes, but it has not spread to staff and workers.

This was the case, as explained earlier, in the privatized corporation when managers were 'encouraged' to opt for individual contracts and leave the union. However, the subsequent cutbacks in management numbers and tougher controls placed on performance levels, carried out under the guise of human resource management and total quality management, made many of the managers we talked to feel that, while in theory the firm had a policy of developing sophisticated human relations, in practice their 'style' had slipped to the traditional mode of non-union cost minimization. This raises the point that to be meaningful management style must be enacted; it must go beyond the crude policy statement of optimistic aspiration and it must also have a capacity to be reproduced. This means that well-meant policy initiatives have to be followed through into action and the action needs to be sustained over a long period so that new patterns of behaviour, akin to culture, become embedded in 'this is the way we do things here'. In the case of the privatized corporation the chairman's initiative in TQM was pronounced a success after a period of eighteen months. What this meant in practice was that productivity per person employed had increased substantially because of the redundancy programme. It was hardly a style change, and attitude surveys revealed deep unease and mounting levels of job-related stress.

Paternalism

The next style box to consider is paternalism (Figure 7.4). Paternalist companies seek to project a welfare-orientated, caring image, couched in terms of stability and order, with employees knowing their place in the firm. Employees are given little, if any,

Paternalism

Paternalist

no collectivism
(unitary)

Fig. 7.4. Paternalist Style

expectation of promotion or job advancement beyond the induction or junior stage. The firm is expected to look after employees and provide reasonable pay and fringe benefits, but in return expects loyalty and compliance. It is usual to find fixed grade structures based on job evaluation and job descriptions. Job training is provided when necessary at the induction stage and when new technology or new working procedures are introduced. All employees affected will receive the training since there is little attempt to differentiate between the performance of individual employees, apart from managers. Communication is exclusively downward and there is a marked tendency for this to be in the form of messages from the chairman and departmental heads, couched in vague but essentially cheering terms, with thanks to the staff for hard work in the previous year. Bad news tends to come in a Captain Oates style of calling (and expecting) sacrifices to be made. This type of communication system can be developed as a cascade system of briefings with messages passed down through the hierarchy in a sequence but with little opportunity for what one cynic called 'splashback'!

These firms also seek to avoid unionization, and conflict, if it occurs, is seen as a failure of communication, or caused by the action of a disaffected employee or a manager making a mistake. Action would be taken against these individuals quickly to avoid the 'trouble' spreading. Often the language used in these circumstances would be akin to the 'rotten apple in the barrel' analogy. Companies pursuing this type of management style were most likely to be found in long-established service and distribution firms like insurance companies, building societies, supermarkets, and department stores. Here there is often a historic division between men and women, with women located in large numbers in the lower grades of the hierarchy. Men have traditionally been seen as the career staff. Although in recent years such firms have sought to change this image and introduce equal pay for work of equal value (after a number of expensive industrial tribunal cases have gone against them) the development of equal treatment in career terms has been more difficult to achieve.

Two of our case studies fitted this style well. In the bank we were told by the managing director of the Scottish subsidiary that they sought to recruit career staff from grammar school boys in Aberdeenshire! Men were expected to be mobile and develop their career from the junior grades by moving from branch to branch. Women were expected to be 'tied' to the home, and to be the second wage-earner in the family, so they were not expected to have the same mobility (see Llewellyn 1981; Crompton and Jones, 1984). The bank was showing concern for family life and family values. This is the best example possible of the definition of paternalism given earlier: 'restricting the freedom of the individual through well-meant regulations'. Similarly, wages for male messengers were higher than those for female secretaries on the grounds that men needed to be paid a family wage. Paternalists often find the modern world confusing, since legislation and the movement for equal opportunity challenge their notion of what the world ought to be like.

In the retail case-study of a large department store in London there were clear status differences between various grades of staff, each with their own canteen, forms of payment, and social standing. Upward communication, in terms of the ability to make suggestions on how tasks and customer service could be improved, was positively discouraged. As in most such status-conscious organizations, the expectation was that a given grade of staff would look up to the grade above and look down on the grade below with temporary and Christmas sales staff seen as the lowest of the low. To know one's place is to know both who is above and who is below. The TV series *Are You Being Served?* captured this perfectly.

Traditional

Moving to the bottom of the individualism scale we reach the box marked traditional (see Figure 7.5). This style is found in companies which see labour as a cost or a factor of production, along with land and capital, as traditional economics puts it. Here labour costs are all-important and efforts are made to minimize costs, labour security, and the expense associated with

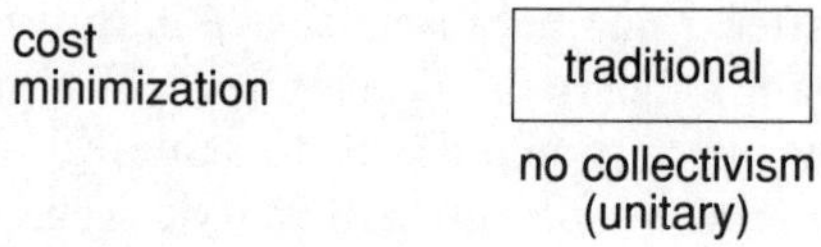

```
    cost
    minimization              traditional

                           no collectivism
                           (unitary)
```

FIG. 7.5. Traditional Style

recruitment, hiring, and training. Employee subordination is assumed to be part of the 'natural order' of the employment relationship, with employees having no recognized role in expressing their views or participating in the management of the firm. The dominant characteristic is one of command-and-control by management, with efforts made to intensify effort and reward only for work done. In some cases there will be extensive reliance on work study based piecework incentive schemes, but in all pay is kept as low as possible. Performance-related pay based on targets is also a possibility. Since such firms expect to have a high labour turnover, except when there is high unemployment, investment in employee development is kept to a minimum and what training there is is provided on the job. These firms fit the type of numerical flexibility proposed by Atkinson (1984) in terms of the use of temporary workers and part-timers, and the reliance of labour turnover to vary the number of employees according to demand. Hours of work are also varied on the same basis, especially for part-timers, who may be asked, and expected, to work for longer than their contractual hours without premium pay. These so called 'peripheral' workers will receive few fringe benefits such as paid holidays, sick pay and pensions. Unions are opposed and every effort is made to avoid membership drives. In any case the high labour turnover makes it difficult for unions to maintain their membership even if they manage to gain a foothold.

Firms with a traditional style are most often found in labour-intensive, low technology industries where the level of skill required is so low that labour can easily be replaced without much training. This would typically be the case in hotel and catering firms, some small to medium-sized clothing companies, and in low-skill franchise operations such as fast-food outlets.

Wong (1992 and 1993) has provided a description of such a firm in the mail order business where changes in consumer demand and taste lead to marked changes in the level of demand

and in the number of returned items (34 million items were dispatched in 1989 with 4 million of them returned within two weeks.) Here the firm uses Saturday workers and especially temporaries, to match labour supply with product demand. Employees who are sick are sacked after two days, and any temporary staff who manage to work for a consecutive period of six months is sacked for a period of a week or more in order to avoid the legal requirements imposed by continuous service, such as protection against unfair dismissal and minimal maternity pay. A further legal threshold is reached at just under two years of service and this is treated similarly. Numerous shift patterns are used to provide working hours over lunch- and dinner-breaks as a means of avoiding overtime rates of pay or paid lunch-breaks. Performance pay is based on speed of packing with employees issued with instructions by the computer for twenty-minute cycles of work. Any employee who wishes to move to a higher cycle time must receive permission from the supervisor. This gives supervisors considerable power and it is they who enforce discipline on the packing floor. Failure to meet the cycle time leads to immediate demotion to a lower rate. In this case base pay is higher than the average for the district to ensure a supply of labour, but additional employment costs are kept as low as possible. As Wong notes, the firm does not describe its employment practices in such harshly exploitative terms, preferring the guise of modernity and progress provided by terms such as 'flexible labour utilization'. Whatever the gloss provided for such management styles the conclusion is inescapable that this is a firm with a traditional style.

It is easy to be critical of such a firm, but their justification of their labour practices is that the price margins in the mail order business are very low and the need to handle large volumes of goods sent out and returned inevitably means there is a need for a large labour force undertaking low-skill, mundane tasks. They claim not to have a choice. There is some strength in this argument and it points to the value of employee protection through labour legislation to avoid 'sweat shops' as they were called in the very early years of the twentieth century. But it is also a reminder to avoid the trap of looking at style purely in terms of good personnel practice without a consideration of the circumstances that face a particular firm.

Not all firms have a capacity to develop resource-based policies and for some of those that attempt to do so the costs will be prohibitive. Style has to fit product market and profitability requirements to a degree, unless a given style is outlawed such that all firms are affected and none can gain competitive advantage through labour exploitation. This may put up the price of a given product (but this is not inevitable since it would become more cost-effective to invest in technology if labour costs were higher—the capital-labour transfer) but there is a welfare case that this should be done for the public good. In the long term the ability of firms to perpetuate a traditional style is to be caught in the cycle of low pay, low productivity. It is very hard for western industrial societies to compete with emerging nations like Thailand where labour costs in 1993 were 0.71 US dollars per hour compared with $12.37 in the UK, $16.4 in the USA, and $24.9 in Germany. Critically therefore this style is more likely to be found in areas where there is no overseas substitution available (like mail order) rather than in manufacturing in free-traded goods.

Bargained Constitutional

The remaining styles take the patterns developed on the individualism scale but add the effect of collectivism. That is to say that our attention now needs to turn to companies which recognize trade unions or allow for forms of employee representation as part of their management style. The largest category is that we call bargained constitutional (see Figure 7.6).

FIG. 7.6. Bargained Constitutional Style

These are firms and organizations which manage employees in much the same way as the paternalists or the traditionalists, but trade unions have been recognized for some time. In terms of choice, managers may believe this to be an unfortunate product of history and they would prefer to manage without unions; it is unlikely that they will value the union presence and will tend to see unions as a constraint. For this reason the opportunity to open a green field site will be used to develop a non-union workforce if that is possible. In the bulk of the company, however, unions are deemed to be inevitable and policies are developed around the need to achieve stability, control, and the institutionalization of conflict. Management prerogatives are defended against the encroachment of unions through the use of highly specific collective agreements, and careful attention is paid to their enforcement and administration at the point of production. Continual efforts are made to restrict the number of issues subject to collective bargaining and joint regulation. Personnel and industrial relations managers play the role of gatekeepers, limiting the involvement of trade unions. The importance of management control is emphasized, with the aim of minimizing or neutralizing union constraints on both operational and strategic management. The decentralization of collective bargaining, as discussed in Chapter 5, is relevant here. The quality of the relationship with trade unions may vary from outright hostility, especially where cost minimization policies are pursued, to more stable systems where unions are accommodated and a regular bargaining relationship is established. The key to the style is the achievement of stability and predictability. It is poor at handling change.

Within Britain at least this has been the dominant style and is the one traditionally expected to be found in the public sector, in mass production and large-batch manufacturing firms, or firms with a large number of employees and a high proportion of semi-skilled and manual workers such as the postal service. We return to consider this style in the later section on changes in style, since this is an area where most change has taken place.

Modern Paternalism

Modern paternalism is the term used to describe firms with a welfare-orientated, paternalist approach to the management of employees, but who nevertheless recognize and value the relationship with trade unions or works councils. Here, stress is placed on the achievement of 'constructive' relationships with trade unions. Extensive information is provided and a network of collective consultative committees established to communicate with

198   *Management Style in the M-form Company*

```
         paternalist          modern
                              paternalist

                           co-operative
```

Fig. 7.7. Modern Paternalist Style

employee representatives. The emphasis is on the needs of the business aiding the management of change within the context of a caring welfare image. Fixed grade structures based on job evaluation are likely to be used to create an impression of a 'felt fair' pay structure, with review bodies to handle job changes and appeals, composed of managers and employee representatives, often in equal number. Quite extensive training is provided to employee representatives to enable them to read company accounts, and understand the principles of marketing, production, and investment so that they can contribute to discussion with senior managers. There is often an extensive use of joint working parties to explore aspects of technical change, multi-skilling, and to handle specific aspects of change. Single-table bargaining and co-operation between unions is positively encouraged, but the danger is that employee representatives get too drawn into management through a process of incorporation and lose touch with their members. This style is especially unstable if the firm changes its approach to employees by moving from paternalism to cost minimization and is thus dependent on the firm maintaining its paternalist labour policies as part of its relationship with unions or works councils. The style is most likely to be found in relatively long-established process firms such as food manufacture, and in the insurance and banking sectors where unions have been recognized for some time.

A good example of this style was evident in one of the divisions in our food company. This division was notable for the degree to which it sought to take the union into its confidence. At the regular joint consultative committee meetings held in a hotel near Heathrow airport, and attended by full-time officers and senior shop stewards from the major plants, management took the union into its confidence, discussing plans and policies openly. For example, five-year forward plans were discussed together with financial and performance data. It was usual for a majority of the divisional directors to be present at the meetings, and a lively discussion often took place on the competitive position of the firm's products compared with that of the major competitors and the approach of the major retailers to shelf space and advertising. A consultant's report on expected excess capacity in the firm was also openly discussed and data provided on relevant productivity figures on a plant-by-plant basis. When the announcement came that a major plant was to close no one was surprised and union opposition was minimal. There was little negotiation on the issue since the style was essentially consultative, that is to say the firm was prepared to encourage a modern paternalist style as long as this provided co-operation and it retained the right of last say. In the adversarial style challenges are expected, and it is therefore preferable not to discuss contentious issues and to preserve the management prerogative. In the co-operative style management still have the prerogative to make decisions but do so under the guise of consultation (McInnes 1985). The union was useful both in providing a source of information on employees' views and in forestalling more serious challenges, the classic 'voice' mechanism described by Freeman and Medoff (1984).

This division had formed the original company and the majority of corporate management staff had built their careers there. This meant that the style developed there was adopted as the preferred style for the whole corporation, but it proved difficult to spread the style to the firm's newly acquired companies, where a form of bargained constitutionalism was in place. This limited the possibility for management in these new divisions to feel it was 'safe' to engage in substantial discussions with the unions and the unions were suspicious of management's intentions when they sought, in a half-hearted fashion, to do so. This illustrates the difficulty of achieving a change in style from bargained constitutionalism where mutual suspicions, typical of an adversarial relationship, limit the possibility of developing a new approach.

Sophisticated Consultative

The final style to consider is the sophisticated consultative (see Figure 7.8). The fundamental approach to the management of employees is very similar, if not identical, to that described earlier in the sophisticated human relations segment. These firms are

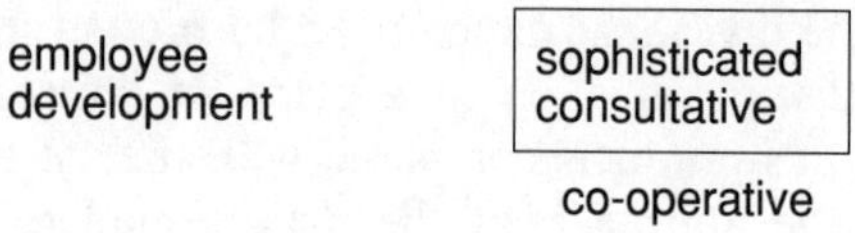

```
employee              sophisticated
development           consultative
                      co-operative
```

Fig. 7.8. Sophisticated Consultative Style

active proponents of total quality management and invest heavily in employees to maximize their contribution to the firm and its productive efficiency and ability to meet customer requirements. Employees, or at least a large part of them, are seen as core elements of the firm's success in the market place. Other employees are more likely to be staff of subcontractors working for example in the canteen or in security. Here management are able to develop a consistent style for their own employees while subcontractors are free to have a different, often less supportive, style for their workers. All the other elements of performance pay, appraisals, careful concern with selection, and competitive teamwork are introduced, together with extensive communication networks and employee attitude surveys. And work remains intensive with little slack time. The expectation is that employees will be committed to the firm and their job and will use their discretion for the benefit of the productive process and the firm in problem-solving, making suggestions, co-operating in quality improvements, and helping others, if need be by working beyond contract. At the most optimistic level these are transformed organizations meeting the dream-like expectations of managers of what the world of work ought to be like. There is a real danger that the rhetoric surrounding such firms is at odds with reality, which might be more mundane. But as a description of strategic intent, if that is what is felt to be required for the firm in order to meet its business objectives and compete with world-class firms, for example from Japan, then this style does drive towards a certain type of policies. European motor manufacturers, for example, have sought to incorporate many of these kinds of labour management principles in recent years through cell manufacturing, total quality management, and team-work (Mueller and Purcell 1992).

The distinctive difference between this style and that of the sophisticated human relations type is the active encouragement of forms of employee representation either via a company council (sometimes linked to a single union agreement as at Nissan in the UK) or through a reformed collective bargaining machinery, usually involving all the recognized unions as in a single-bargaining unit. The attempt is made to build 'constructive' relationships with trade unions and incorporate them into the organizational fabric, often as part of company councils. These are similar to works councils in Germany, but without the supportive legal regulations. The lack of legal support can be seen to limit the effectiveness of company councils which can be restricted to a purely advisory role, as seen in the title of the relevant body in Toyota in its new plant in the UK. This is called the Toyota Members Advisory Board and the recognized union has no guaranteed membership of the council. However, the existence of a form of final offer, or pendulum arbitration is seen to provide a basis for difficult issues to be resolved externally, in something akin to a weak form of labour court, but again without extensive legal rights. Much depends on the way in which management seeks to make such company councils work. At their best there will be wide-ranging discussions, with extensive information provided on a whole range of decisions and plans with the aim of gaining consensus while recognizing that such bodies are not engaged in joint regulation, in the proper meaning of the term, but in strategic consultation. Management, in the end, will decide. This style is much favoured by the large Japanese plants in the UK (but not in the USA where a non-union approach is more likely) and by some green field site operations where the decision is taken to recognize trade unions, if only to forestall union recruitment campaigns later. A good example is the Pirelli General plant in Wales (Yeandle and Clarke 1989).

CHANGING MANAGEMENT STYLES

The sophisticated consultative style is seen as the model type for some companies to aim for from the bargained constitutional box if the strategy includes the continued recognition of trade unions. This was clearly the case in the Rover Motor Company in its attempts at modernization in the early 1990s. The essence of the company's approach is shown in a company document called 'Rover Tomorrow—The New Deal' with the sub-title 'We need a workforce distinguished only by individuals'/teams' contribution to the company'. The document is reproduced in Appendix 1. Paragraph 14 is a good example of the way in which relations with trade unions are expected to fit the new deal: 'consultation with representatives of recognized trade unions will be enhanced to ensure maximum understanding of company performance, competitive practices and standards, product and company plans and all areas of activity affecting the company and its employees.'

The then personnel director of Rover, Rob Meakin, explained in an earlier article that he

is presenting the new deal to unions as one alternative in a stark choice that has been forced on British car makers by circumstance. We have seen the best car makers in the world move out of Japan. We saw them take their approach and products to the US and could plot the closure of US manufacturers as Japanese plants opened there. Now three of those companies are setting up in the UK and they will not be bringing any customers with them. Do we want to take a chance and hope the storm passes by, or take the opportunity to face up to it?. . . Rover now faces the imminent arrival of Toyota and Honda to join the existing Nissan plant. Productivity levels are much higher than at Rover. What we have got here, and why the partnership with trade unions is important, is an opportunity for a traditional, multi-union engineering company with 85 years' history to demonstrate it can organize and operate at levels of productivity that compete with the best in the world—even if the best are non-unionized.   (Personnel, 22 Oct. 1991, p. 25)

The Rover case and many others like it, for example Thorn Lighting (Industrial Relations Review and Report 1992: 12–15) illustrates the way in which attempts have been made in manufacturing to shift the style from an adversarial position to one of sophisticated consultative.

This raises the question of the direction of change and the problem of maintaining an adversarial style with trade unions while seeking to gain high individualism based on team-work, multi-skilling, and task flexibility. The difficulty is shown in Figure 7.9 as the second of the unstable conditions. Is it possible to maintain adversarial relationships with trade unions while pursuing high individualist commitment policies with employees? The problem here is that the company recognizes that it needs to

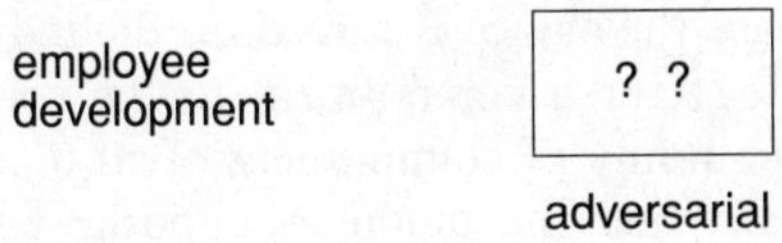

employee development	? ?
	adversarial

FIG. 7.9. Unstable Style II

develop world-class manufacturing standards and sees that this will require a change in the way it manages and motivates employees; the company will have to introduce the range of policies described earlier in the section on sophisticated human relations. Given the adversarial nature of the relationship with the union representatives, who are also in most cases shop stewards and thus employees of the firm in close touch with the shop floor, it is necessary to negotiate the proposals. However, this means that management will believe that each change has to be costed and haggled over in a long-drawn-out, expensive process. The unions are very likely in these circumstances to see the proposals for direct communication with employees, team-work, and performance-related pay as a direct attack on their representational rights and abilities. For example, they will be likely to emphasize that they should be the 'single channel of communication' with the workforce, as the unions did at the time of proposals for worker–directors in the Bullock Report of 1976. The problem is that in adversarial conditions both parties expect the worst of each other. This either leads to attempts to bypass, or leaves both sides doing nothing for fear of stirring up trouble. And if action is taken this is often done in anticipation of a hostile reaction—usually a self-fulfilling prophecy. One traditional approach to this problem is the use of productivity bargaining, but the history of this type of bargained change has generally been lamentable, with companies paying for work practice changes before they have been implemented, as we described in our discussion of the privatized corporation. In some cases managements have paid for changes many times over (Ahlstrand 1990).

This problem box is unstable for another reason. If the company is able to gain employee co-operation and introduce sophisticated employee resource policies, and if these are valued by employees in terms of commitment and motivation then it becomes more difficult for unions to perpetuate an adversarial stance. In a sense the choice is forced on the individual employee of whether to opt for a commitment to the company or to the union. This dichotomy of commitment often unnecessarily polarizes the company and the union as opposite ends of one spectrum. In fact employees can show commitment to both the union and the company. It does depend, however, on the behaviour of each party to the other. The more adversarial the relationship and the greater the extent to which the management adopts cost minimization policies the more likely it is that commitment, if it exists at all beyond general apathy, will be to the union. Similarly, if the employment practices are based on high individualism and employee resource policies and the union seeks to maintain an adversarial, oppositional stance, then it is likely that commitment will be more in favour of the company.

In these circumstances the unions' worst fears are likely to be realized as workers cease to support the union, fail to attend meetings, vote against union proposals in ballots, and eventually leave the union altogether most likely by not paying subscriptions. It will also be difficult for the union to recruit new members from among young, newly appointed staff. Eventually the style changes to a non-union, sophisticated human resource model. Once membership has fallen to well below half of the workforce, management will eventually decide that it is no longer worth maintaining recognition and will make a judgement that the union does not have the strength to resist de-recognition, as happened in Unipart in 1992, and in our electronics case-study. Here union membership was just over 20 per cent at the time of the de-recognition. The company argued that to continue with recognition was to perpetuate the disenfranchisement of a large majority of its staff. They created a directly elected company council in place of the union.

In effect, therefore, the difficult choice facing companies in the bargained constitutional box, and in a different sense facing the union too, is which direction to choose once it is recognized that there has to be a change towards greater individualism employee development-type policies, as in Rover. Should the target be top left (sophisticated human relations and a non-union firm), or based on the growth of co-operative relations with unions and the emergence of a sophisticated consultative style? Some very large companies such as British Rail and the Royal Mail attempted the former in the mid- to late 1980s when it was fashionable to take on the unions, and in both cases they failed. Even in our privatized corporation, union membership among managers, who were encouraged to opt for individual contracts, remains surprisingly high. In the case of the predecessor to Rover, British Leyland, the 'Blue Newspaper' of 1980 was widely seen as an attempt to drive unions out, and certainly to emasculate shop-steward power (Willman and Winch 1985). Later, emphasis was placed on 'constitutional' trade unionism, which meant limited negotiating power and extensive consultative rights. Thus, the first reaction in these companies was to move to the non-union model, but this changed when significant employee opposition was encountered and, as they now say in Royal Mail, a form of 'strategic partnership' came to be encouraged.

In these cases the option of moving upwards and to the left in the matrix was closed, except for managerial staff, and the only route left was via the development of co-operation with the unions. In some cases this was an option forced on the unions. Many writers on human resource management in a unionized context note that it is not possible to sustain adversarial relationships and that unions can be compelled to change their stance, or find that they are bypassed in the change process, left with the residual role of negotiating minimum pay awards once a year, or even once every three years. Storey quotes a senior personnel manager at Rover to illustrate this point: 'The unions were invited to the party but they didn't want to come. So the party went ahead without them' (1993: 544). He goes on to note an extra twist in the change process referring to the marginal role of personnel:

Equally, it must be remembered that the internal politics of change were such that the people issuing the 'invitations' were in any case rarely the main drivers of change. The major change programmes were frequently devised outside the specialist personnel industrial relations function. In consequence, by the time this branch of management got a hold on the change package its shape was pretty well settled. As personnel were still widely regarded as the chief mediators with unions, this inevitably meant late involvement by the unions.

Faced with this, both unions and personnel have to adapt to the new forms of employee development and a co-operative, more consultative, stance, or be driven out by being rendered irrelevant.

Once unions adapt to human resource management they can play a wider but less central role as opinion leaders, exercising the voice mechanisms. For example the Thorn Lighting case shows how attempts to foster team-work ran into difficulty because of low trust between management and the workforce. They tried to initiate team-building courses, but found it difficult to persuade employees to attend:

> The company decided to approach the shop stewards, with whom it described itself as having a good relationship, to discuss ways round this problem. Employees were informed that 'they had nothing to fear' from the programme, and a small number eventually agreed to take part. . . . Now some 850 employees have taken part in batches of 18 delegates per weekend' (IRRR 1992: 14–15). A similar shift in union attitudes is reported in Rover (Industrial Relations Review and Report 1993: 5–7).

Not all companies have chosen to go in this direction; some have preferred to develop non-union sites as part of their green field strategies. Here the attempt is often made to begin with an employee resource policy such as soft human resource management and to avoid trade unions, even if unions are recognized in the traditional heart of the company. The aim is often to transplant the best practices developed in the green field site to the original sites and 'force' a form of sophisticated consultation or, by progressive investment and expansion in new sites, eventually to close the old. As we noted earlier, M-form companies have a capacity to reward or punish through their investment policy.

The issue for the company in choosing the most appropriate style for its circumstances is whether in the long run it wishes to avoid unions, or seeks a transformed relationship with them. This presupposes a degree of strategic thinking which Storey, in his study of fifteen large British companies, generally found lacking. 'Across a variety of approaches, one startling fact stood out: while the old-style industrial relations "firefighting" was disavowed and even scorned, there was hardly an instance where anything approaching a "strategic" stance towards unions and industrial relations could be readily discerned as having taken its place' (1993: 529–49). Perhaps the only group who do think strategically in this way are inward investors, who often employ

Management Style in the M-form Company 207

consultants to develop best practice for companies developing green field sites. It is important to note, as Newell (1991) makes clear, that companies opening new sites in development areas in the 1950s and 1960s did not think strategically either about industrial relations or work and employment relations in terms of job design or personnel management. They sought, in the main, to replicate existing practices. There is more evidence of strategic thinking in human resource management than there was, but there is still a long way to go before it is recognized as a key part of strategic management generally. We explored reasons for this in Chapter 4.

The implication so far has been that companies thinking strategically about change in employee relations in the 1990s are all concerned with seeking to move upwards towards employee resourcing policies. This is where, rhetorically speaking, the trendy new imagery of human resource management is to be found. We noted earlier how it is unlikely that more than a small number of companies will be able to meet the requirements of this type, and these will be companies where a high proportion of skilled, or multi-skilled, employees will be required, often associated with an enhanced requirement for diagnostic skills as technology advances. Those jobs that are deskilled tend to disappear altogether either directly or via subcontracting, and a smaller number of functionally flexible employees are required. This patently does not fit the requirements of all companies. In particular it is possible to identify a clear trend towards a downward shift in the style matrix from paternalism to cost minimization. This is clearly associated with subcontracting of, say, canteen services, as in our privatized corporation, with the contractors cutting wages, and more especially staffing levels and fringe benefits, in order to gain the contract under competitive tendering. The same is true of contract cleaning, direct labour services in local authorities, and competitive tendering in hospital ancillary services (Bach 1989; Ascher 1987).

Indeed more generally it can be suggested that, as markets impact more on organizations because of privatization, competitive tendering requirements, or because of the collapse of cartel arrangements and deregulation in banking and insurance, with the associated loss of profitability, firms are *required* to move out of paternalism, either towards high added-value labour policies of employee development (for an Australian example see Mathews (1991) study of Colonial Mutual Life), or to cost minimization. The implication is that we are seeing a more radical bifurcation of employee relations. In terms of the style matrix this is either downwards to cost minimization or upwards to employee development, either to non-union positions, as in the USA, or towards the top right of the matrix into the sophisticated consultative style favoured in continental Europe in countries like Germany and Sweden, where employee participation through works councils and employee director schemes is accepted as 'normal'. The potential movements between styles are illustrated in Figure 7.10. The figures relate to a summary of the type of changes that can be expected. It will be observed that both paternalist and adversarial positions are likely to reduce in number and importance.

If the Social Chapter of the European Community is eventually enforced, especially the section concerned with the creation of European works councils (Hall 1992), then legislative pressure will force companies towards the sophisticated consultative style. This in turn will require M-form companies to consider their management style afresh. In our staple products company we reported earlier that, while they allowed for recognition of trade unions at plant level and adopted a policy of parochialism for unions by restricting involvement to plant affairs, they 'prided themselves on never having had a union official crossing the threshold of the corporate office' (in London). The European works councils would require them to break this rule, at least in respect of their continental European plants in the Netherlands and Germany. However this 'threat' is a long way off and the personnel director noted that the original proposal for industrial democracy in the EC was made in 1973 and has yet to reach the statute book. In line with one of the roles of the corporate personnel department suggested in Chapter 5, the director is lobbying in Brussels to ensure that the latest proposal is also stillborn.

When we searched for coherent management style statements in our sample of M-form companies, we were unable to find much evidence of anything beyond generalized banality, certainly nothing that constituted a considered policy. This led us to ask whether it was possible for such organizations to develop a well thought-out set of principles for the management of employees.

Management Style in the M-form Company 209

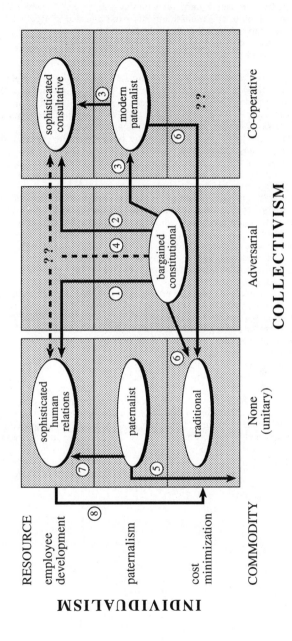

FIG. 7.10. Movements in Management Style in Employee Relations

[1] Employees encouraged to sign new individual contracts and union recognition for collective bargaining purposes withdrawn. New work practices, recruitment, selection, appraisal, and training initiatives implemented in accordance with 'soft' HRM and employee development policies.

[2] Co-operative, consultative relations with trade unions/works councils/company councils developed alongside the introduction of employee development-type policies.

[3] Co-operative, consultative relations initiated with unions as a prelude to subsequent initiatives in employee development. The new relationship with unions is often triggered by a crisis in the competitive position of the firm such that choice is forced. Competitive analysis reveals that the strength of the major firms in the market is based in part on employee development policies.

[4] Unstable conditions exist as unions are bypassed in the change programme to introduce employee development policies. Subsequently either union membership declines and recognition is withdrawn, or both unions and management 'learn' to modify their behaviour to each other to emphasize partnership, or initiatives fail and the bargained constitutional pattern is reinforced.

[5] Growing competition, falling profit margins, and declining market share force a reappraisal of employment policies leading to the introduction of cost minimization, reduction in job security, and reduced employee benefits. Or new entrants to the market base their competitive advantage on cheaper labour.

[6] A new tough regime is introduced often triggered by a change in ownership, competitive tendering, subcontracting or acquisition. Union recognition is withdrawn and cost minimization policies reinforced, with employees working under worse conditions.

[7] Emphasis placed on employee productivity achieved through employee development policies and technical change based on realization that it is desirable to encourage employees to use diagnostic skills and their knowledge for the benefit of the business and to satisfy customers. Usually associated with reduction in number employed and thus rapid, early rises in productivity. The difficulty is to sustain this.

[8] Rapidly falling market share in a depressed or mature market leading to substantial loss of profitability forces a major reappraisal of employment policies. These are often justified on the grounds of temporary expediency but, once implemented, are difficult to escape from. A change in top management is often a precursor to the abandonment of the employee development policies, and substantial cuts in employee investment. Often triggered by the arrival of low-cost entrants to the market and a slow response by the firm, leading to crisis and draconian action.

In the end we concluded that the greater the extent to which the company has grown by acquisition, and has divisionalized, diversified, and decentralized, the less capable it is of developing integrated management styles towards the way employees are to be managed across the whole corporation. They are more likely to rest responsibility for this in operational management. The problem, as we have noted earlier, is that if these firms simultaneously require short-term pay back on investment and impose strict financial targets this drives out the possibility that operational units can themselves pursue long-term employee development strategies.

The only exception to this general rule is where the division is operating in a market where the lead companies in a global sense have achieved high productivity levels through the adoption of such policies. This would be the case in Rover, a division of British Aerospace, where the cost, nationally, of failure would be enormous. Their problem is to survive the recession and build market share in the key segments of the market. Influences on style choice come from the corporate headquarters and its guiding principles, the industry standards set by the lead players, and the cultural influences of society in general, reflected in the labour market and the actions of government. Within these general influences the more precise choices are likely to be made at the business level linked to the product market, choices on technology, and skill level. But operational management look to the corporate office for signals or guidance on policy and style. The key is strategic planning in an integrated manner.

SUMMARY: MANAGEMENT STYLE AS A STRATEGIC TOOL

Management style means the preferred way of managing employees. It is composed of two dimensions, individualism and collectivism. *Individualism* centres on the types of policies and practices concerned with the way in which jobs are designed, people rewarded, contracts established, and training and development of employees undertaken. At the high end of the individualism scale are those managements which seek to invest in people and focus on each person's contribution, capability and competencies. This implies a set of employee resource and development policies often associated with internal labour markets, human capital formation, and organization-specific employment policies. This is most likely to be found in companies which have relatively high ratios of capital to labour, require skilled, diagnostic work to be done, often involving team-working, and operate in a market where there is competition, often on a global scale, based on quality, reliability, and innovation.

At the low individualism end of the scale are companies where labour is seen as a factor of production, or a commodity, which is hired from the external labour market as and when required. Little investment is placed in employees and numerical flexibility is preferred, achieved through variations in numbers employed, hours worked, or work contracted out. The prime motivation underlying employee relations is cost minimization. Managements which utilize this style more often employ relatively large numbers, or a high proportion, of low-skilled employees (often women and people from racial minority groups), labour costs constitute a high proportion of total costs, or of total variable and controllable costs, and product demand conditions are often unstable or variable.

In between these two extremes, but a distinctive category in its own right is paternalism. This is where the long-standing approach to the management of people at work is based on assumptions of job security, loyalty, and an expectation that, beyond career-grade staff (usually male), a person, once in post, is likely to continue in the job until resignation or retirement.

The *collectivism* dimension of the style scale describes the extent to which the company and its management seeks to avoid or recognize trade unions or other forms of employee representation, and the nature of that relationship, if recognition is granted or maintained. Here the choice is whether to seek to build a form of partnership and emphasize joint problem-solving with extensive consultation on aspects of the business and work organization, or to limit the interaction to a form of adversarial behaviour. This latter is the classic or expected form of union–management relations in the UK. It is coming to be challenged as union power recedes and managements seek to change their style of employment relations. A key question is what type of style will emerge over the next decade.

Management Style in the M-form Company

Management style was developed and defined in this chapter by examining the interaction of the two dimensions of individualism and collectivism. Six potential styles were described, together with two theoretical possibilities which, it was argued, were inherently unstable. How can adversarial collective bargaining and antagonistic union management relations exist alongside employment policies based on 'empowerment', team-work, training, and high investment in employees? Similarly it was felt to be difficult for a union or works council to enter into a collaborative and co-operative relationship with a management who were simultaneously operating a labour policy of cost minimization and exploitation.

The chapter has sought to describe the six styles in some detail and provide the justifications for each. These were sophisticated human relations, paternalist and traditional styles in non-union companies, and bargained constitutional, modern paternalist, and sophisticated consultative styles in unionized firms or where alternative forms of employee representation are allowed for. Described in this way they can all too easily be seen as static descriptions of current or past practices. There may be some value in this in its own right, but the more powerful purpose of the matrix is to provide a means to plot changes in style over time, and, more importantly to help determine future styles to fit the needs of the firm in its changing environment. Some illustrations were given towards the end of the discussion in plotting changes that have occurred in some companies and a summary of the type of possible movements was given in Figure 7.10.

The underlying proposition is that the two central positions of paternalism and bargaining constitutionalism are in the process of being squeezed out, since both are dependent on stability, order, and predictability. One of the personnel directors in our case studies said that, for him, the driving force for change in his company, and in the way he now thought strategically about employee relations, was 'the realization that there is no calm water around the next headland'. It was no longer a case of belt-tightening or taking short-term action to weather a storm but a need to cope with chaos, build for flexibility, and find ways of integrating employment relations into the fabric of the organization.

In some cases this has meant rapid change to reduce overheads and intensify work, subcontracting out services and minimizing costs wherever possible. This can also be associated with union de-recognition or a marginalization of the union role and reduction of the scope of collective bargaining—a shift to the traditional style. In other cases, more often found in advanced manufacturing, the shift in style was planned around the need to introduce cell manufacturing and total quality management. It involved changes in job design, team-work, new forms of payment systems, employee involvement, enhanced training, and a considerable change in the role of line management toward what is grandiosely called 'transformational leadership'. It often also meant a reduction in the number of employees and improvements in efficiency such that there was less spare time. In these cases the style change was upwards towards the employee development patterns. Usually it was associated with a change in the relationship with trade unions, or the creation of new forms of employee representation such as company councils. While the dialogue was extensive it was based more on consultation and information exchange and less on bargaining (the sophisticated consultative style). In a few cases the union role disappeared and the preferred style was one of sophisticated human resources. It may be that if union membership continues to decline, especially if unions are unable to replace old members from among the younger workers that this style will grow further, unless aspects of the European Social Chapter come to be implemented across the Community, including the UK.

We have sought in this book to provide an integrated approach to theory and practice based on our research and our own experience as teachers of managers and management students, and as consultants. The style matrix, over the ten years that it has slowly evolved, has proved to be extraordinarily useful. It continues to evolve. In a practical setting, the matrix is best used as a diagnostic tool, as an aid to mapping change.

For a given company, union or group of employees the key questions that have to be asked are:

1. What is your current style? How do you evaluate its strengths and weaknesses? Why does (or did) this style exist?
2. Are there differences in style in different parts of your organization and for different types of employees? Why might this be

so? Are differences a weakness or a strength? If there are differences, how do you manage the boundary between them?

3. Is there a corporate style statement? Should there be one? Who should write, or articulate, it? How can it be enforced or made meaningful? What is the point of the banal style statements so commonly found?

4. What style will be required in five years' time? Why? What factors in the business, political, economic, social, and legal environment are tending to push the company towards a change in style?

5. If you identify a need to change style (and in our experience every company does), how do you intend to plan for the changes and implement them? The need here is to consider action on both axes of the matrix, individualism and collectivism.

6. If it is your strategic intention to change the way you manage employment relations, what action is needed first to design the changes and then implement and implant them such that the culture of the organization is shifted and the changes take root and have the capacity to reproduce and continue?

7. And finally, what are the key resistance points to change and how do you intend to change them or learn to live within their constraining influence? [Interestingly here many managers mention short-termism, the baleful influence of the capital market, and the 'problem' of middle management accepting change.]

We have left style to the end of this book since it is the most fundamental of all of the choices that face corporations in the management of employee relations. Decisions on the structure and function of the corporate personnel office flow from it. The structuring of internal labour markets and the location of collective bargaining will be deeply influenced by decisions on management style. Remarkably few organizations seriously consider questions of management styles and the most effective way to manage their employees. Strategic thinking in this area has been rare, but the greater the pace of change in all aspects of economic life, the more companies are forced to make choices. It is always possible to choose to do nothing, or to be reactive but the more we look at successful global companies the more it is evident that there is a rationale for designing human resource management to add to competitive advantage and economic success.

APPENDIX 1

Rover Tomorrow—The New Deal

We need a workforce distinguished only by individuals'/teams' contribution to the company

1. Rover will be a single-status company. We are all employees, and the only distinction is the contribution we make. All remaining distinctions between 'staff and hourly-paid' status will be ended:
   - Clocking will be progressively phased out for all employees.
   - A single-status sick pay scheme will be progressively introduced.
   - All employees will be invited to participate in a regular health check provided by the company.
   - Everyone working within the company with the exception of external-facing activities will wear company workwear. We will all wear appropriate safety apparel: hearing protection, safety glasses etc.
   - Single-status catering will apply throughout the company.
   - Production-related employees will progressively have the opportunity to take part of their annual holiday entitlement outside shutdown periods.
   - Every employee will be paid by credit transfer.
   - No employee will be laid off by the company. In the event of a problem which disrupts production, all employees will be engaged in worthwhile activities and be required to co-operate with efforts to maintain productive output.
   - The minimum notice period for all employees will be one month.

2. Continuous improvements will be a requirement for everyone . . .
Work planning/changes in production schedules will be carried out in consultation with the teams involved to ensure the most effective route and use of all resources.
   - The teams will consider all alternative ways of satisfying customer demand.
   - If overtime working is required it will be allocated fairly based on skills/numbers required with no restrictive practices applied or sought.

Appendix 1: Rover Tomorrow

3. Employees will be expected to be flexible subject to their ability to do the job, after training if necessary, and subject to safe working practices being observed. Every employee will have unrestricted access to the use of company tools and equipment necessary for them to make their contribution.

4. There will be maximum devolution of authority and accountability to the employees actually doing the job. Teams will be responsible for:
- quality of work;
- routine maintenance;
- routine housekeeping/waste materials disposal;
- involvement in plant/office layout and equipment;
- process improvements;
- cost reduction;
- control of consumable tools and materials;
- work allocation;
- job rotation;
- training each other; and
- material control.

5. It is our intention to establish a single-grade structure for all our people. The current five-grade 'hourly-paid' and six-grade staff structure will be progressively replaced by a scheme with a reduced number of occupational classifications. Each level will have a salary band which will be achievable by everyone through skill acquisition.

6. Productivity bonus schemes will be progressively phased out. All of us will participate in a bonus scheme related directly to the performance of the company. Qualification for the bonus will be attendance-related . . .

8. Employees who want to work for Rover will be able to stay with Rover. Necessary reductions in manpower will be achieved in future, with the co-operation of all employees, through retraining and redeployment, natural wastage, voluntary severance and early retirement programmes. . . .

10. All of us will participate in identifying training needs and giving and receiving training to improve skills/knowledge and to continuously improve the processes on which we work. . . .

14. Consultation with representatives of recognised trade unions will be enhanced to ensure maximum understanding of company performance, competitive practices and standards, product and company plans and all areas of activity affecting the company and its employees. Twice-yearly reviews of company performance and outlook will take place with national officials and joint negotiating committee employee representatives.

- Training of employee representatives will be developed and strengthened on a joint trade union/company basis to ensure representatives and managers have the fullest understanding of agreements, company and trade union philosophy and objectives, and the skills necessary to fulfil their responsibilities.
- All employee representatives will be encouraged to become fully involved in the development of safe working practices and plant layouts, improvement of medical, catering and all other employee facilities, employee training, understanding of pensions and other benefits, etc.

15. In the event of any grievance or dispute which any employee or group of employees may have, the full company/trade union procedure will be used to resolve the problem. In the unlikely event of any grievance or dispute not being resolved in this manner, if both parties agree, it will be referred to arbitration—the outcome of which will be binding on both parties. There will be no disputes outside this procedure.

Industrial Relations Review and Report Employment Trends 514 June 1992.

APPENDIX 2

Note on the Main Statistical Sources Used

The book draws on more than our case research and earlier literature. In particular, reference is made to two different types of surveys which complement each other. In 1984 one of us was invited to join a research team at the University of Warwick to conduct a large-scale survey of major companies in the UK. Interviews were undertaken with senior personnel managers in the corporate office of 106 enterprises employing 1,000 or more people in the UK. Matched questionnaires were administered to similar personnel people in 97 divisional offices and 175 establishments. This survey is known as the first company-level industrial relations survey with the ugly acronym, CLIRS 1. The main publication, *Beyond the Workplace: Managing Industrial Relations in the Multi-Establishment Enterprise* was published in 1987 (Marginson, Edwards, Martin, Purcell, and Sisson). In 1992 the second company-level industrial relations survey was undertaken (CLIRS 2). This involved interviews in 176 corporate offices of firms with 1,000 or more employees in the UK. It is a representative sample. In this survey there was a matched interview in each case administered to the senior personnel person and to a senior finance executive. The initial analysis was published as a Warwick Industrial Relations paper (no. 45) in November 1993 (Marginson, Armstrong, Edwards, and Purcell). The different sample design of the two CLIRS surveys makes it impossible to draw statistically valid comparisons between 1985 and 1992 even if the temptation is to look for trends in the data.

CLIRS 1 and 2 were designed partly in response to the influential workplace industrial relations surveys. The first of these was undertaken in 1980 and is known not surprisingly as WIRS 1. The second survey was in the field in 1984 with the results published in 1986 (Millward and Stevens). WIRS 3, the third workplace industrial relations survey, was conducted in 1990 with the first book of survey published in 1992 (Millward, Stevens, Smart, and Hawes). The strength of the WIRS surveys is their coverage of workplaces or establishments across the whole of the economy. Only the smallest workplaces with less than 25 people are excluded. The fact that there are three surveys means that a form of time series data can be developed from 1980 to 1990, a momentous decade. The surveys are referred to in the text as appropriate. The limitation of the WIRS surveys is that they focus on the place of work. The whole thrust of this book, and the purpose of the company-level surveys, is to show that it is the policies developed at the apex of the corporation which play a dominant role in determining the type of management activity in industrial relations and human resource management lower down in the office and factory. In particular the combination of detailed case studies undertaken over a period of years combined with powerful and unique statistical data from detailed interviews allows us to draw a picture of strategies and structures in the multi-divisional firm and the links with policies and practices in human resource management.

REFERENCES

ACAS (1983), *Collective Bargaining in Britain: The Extent and Level*, Discussion Paper no. 2 London: ACAS.
ACAS (1993), *Annual Report 1992*, London: ACAS.
AHLSTRAND, B. W. (1990), *The Quest for Productivity: A Case Study of Fawley after Flanders*, Cambridge: Cambridge University Press.
—— and PURCELL, J. (1988), 'Employee Relations Strategy in the Multi-Divisional Company', *Personnel Review*, 17/3: 3–11.
ALLEN, L. (1958), *Management and Organization*, New York: McGraw-Hill.
ALLEN, P. T., and STEPHENSON, G. M. (1985), 'The Relationship of Inter-Group Understanding and Inter-Party Friction in Industry', *British Journal of Industrial Relations*, 23/2:, 203–13.
ALPANDER, G. C., and BOTTER, C. H. (1981), 'An Integrated Model of Strategic Human Resource Planning and Utilization', *Human Resource Management*, 1: 189–203.
ANDREWS, K. R. (1980), *The Concept of Corporate Strategy*, Homewood, Ill.: Irwin.
ARGENTI, J. (1974), *Systematic Corporate Planning*, Sunbury-on-Thames: Nelson.
ARMSTRONG, P. (1984), 'Competition Between the Organizational Professions and the Evolution of Management Control Strategies', in K. Thompson (ed.), *Work, Employment and Unemployment: Perspectives on Work and Society*, Milton Keynes: Open University Press.
ASCHER, K. (1987), *The Politics of Privatization*, London: Macmillan.
ATKINSON, J. (1984), 'Manpower Strategies for Flexible Organizations', *Personnel Management*, Aug.: 28–31.
——, and MEAGER, N. (1986), *Changing Patterns of Work: How Companies Introduce Flexibility to Meet New Needs*, Falmer: IMS/OECD.
BACH, S. (1989), *Too High a Price to Pay? A Study of Competitive Tendering for Domestic Services in the NHS*, Warwick Papers in Industrial Relations, no. 22, Warwick: University of Warwick, Industrial Relations Research Unit.
BAMBERGER, P., and PHILLIPS, B. (1991), 'Organizational Environment and Business Strategy: Parallel versus Conflicting Influences on Human Resource Strategy in the Pharmaceutical Industry', *Human Resource Management*, 30: 153–82.

BOXALL, P. (1992), 'Strategic Human Resource Management: Beginnings of a New Theoretical Sophistication?', *Human Resource Management Journal*, 3: 60–79.

—— (1993), 'The Significance of Human Resource Management: a Reconsideration of the Evidence', *International Journal of Human Resource Management*, 4/3.

BULLOCK, LORD (1977), *Report of the Committee of Inquiry on Industrial Democracy*, Cmnd. 6706, London: HMSO.

BUTLER, R., DAVIES, L., PIKE, R., and SHARP, J. (1987), 'Strategic Investment Decision-making: Complexities, Politics and Processing', paper given at the British Academy of Management conference, University of Warwick, Sept. 1987.

CALLUS, R. (1991), 'The Making of the Australian Workplace Industrial Relations Survey', *Journal of Industrial Relations*, 33: 450–67.

CAPPELLI, P., and SINGH, H. (1992), 'Integrating Strategic Human Resource and Strategic Managements', in D. Lewin, O. Michell, and P. Shever (eds.), *Research Frontiers in Industrial Relations and Human Resources*, Madison: IRRA.

CBI (1988), *The Structure and Processes of Pay Determination in the Private Sector 1970–1986*, London: CBI.

CHANDLER, A. (1962), *Strategy and Structure*, Cambridge, Mass.: MIT Press.

—— (1976), 'The Development of Modern Management Structure in the US and UK', in L. Hannah (ed.), *Management Strategy and Business Development*, London: Macmillan.

CHANNON, D. F. (1982), 'Industrial Structure', *Long Range Planning*, 15/5: 78–93.

CHRISTIANSEN, E. T. (1987), 'Challenges in the Management of Diversified Companies: The Changing Face of Corporate Labor Relations', *Human Resource Management*, 26/3: 363–83.

CLEGG, S. (1990), *Modern Organizations: Organization Studies in the Post-Modern World*, London: Sage.

CRAFTS, N. (1991), 'Reversing Relative Economic Decline: The 1980s in Historic Perspective', *Oxford Review of Economic Policy*, 7/3: 81–97.

CRESSEY, P., ELDRIDGE, J., and MACINNES, J. (1985), *Just Managing: Authority and Democracy in Industry*, Milton Keynes: Open University Press.

CROMPTON, R., and JONES, G. (1984), *White-Collar Proletariat*, London: Macmillan.

DAFT, R. L. (1989), *Organizational Theory and Design*, St Paul: West Publishing Co.

DEERY, S., and PURCELL, J. (1989), 'Strategic Choices in Industrial Relations Management in Large Organizations', *Journal of Industrial Relations*, 31/4: 459–77.

DONOVAN, LORD (1967) *Royal Commission on Trade Unions and Employers' Associations, 1965–1968*, Cmnd. 3623, London: HMSO.
DORE, R. (1989), 'Where are We Now?: Musings of an Evolutionist', *Work, Employment and Society*, 3: 425–46.
DYER, L. (1984), 'Studying Human Resource Strategy', *Industrial Relations*, 23/2: 156–69.
—— (1988), *Human Resource Management: Evolving Roles and Responsibilities*, Washington, DC: Bureau of National Affairs.
EDWARDS, P. K. (1987a), 'Factory Managers: their Role in Personnel Management and their Place in the Company, *Journal of Management Studies*, 24/5: 479–501.
—— (1987b), *Managing the Factory*, Oxford: Blackwell.
ELGER, T. (1990), 'Technical Innovation and Work Reorganization in British Manufacturing in the 1980s', *Work, Employment and Society*, May (additional special issue): 67–102.
FERNER, A., and HYMAN, R. E. (1992), *Industrial Relations in Europe*, Oxford: Blackwell.
FLIGSTEIN, N. (1985), 'The Spread of the Multi Divisional Form Among Large Firms, 1919–1979', *American Sociological Review*, 50. June.
FORTUNE. (1988), 'Corporate Strategy for the 1990s', *Fortune*, Feb. issue.
FOULKES, F. K. (1980), *'Personnel Policies in Large Non-Union Companies*, Englewood Cliffs, NJ: Prentice Hall.
FOX, A. (1966), *Industrial Sociology and Industrial Relations*, Royal Commission On Trade Unions And Employers' Associations, Research Paper 3. London: HMSO.
FRANCO, L. G. (1974), 'The Move toward a Multidivisional Structure in European Organizations', *Administrative Science Quarterly*, 19: 493–506.
FREEMAN, R. B., and MEDOFF, J. L. (1984), *What Do Unions Do?*, New York: Basic Books.
GALBRAITH, J. K. (1972), *The New Industrial State*, London: Andre Deutsch.
GOOLD, M., and CAMPBELL, A. (1987), *Strategies and Styles: The Role of the Centre in Managing Diversified Corporations*, Oxford: Blackwell.
GOSPEL, H. F. (1992), *Markets, Firms and the Management of Labour in Modern Britain*, Cambridge: Cambridge University Press.
GOWLER, D., and LEGGE, K. (1981), 'Groups that Provide Specialist Services', in R. Payne, and C. Cooper (eds.), *Groups at Work*, London: Wiley.
GRANT, R. M. (1991), *Contemporary Strategy Analysis*, Oxford: Blackwell.
GRINYER, P., YASAI-ARDEKANI, M., and AL-BAZZAZ, S. (1980), 'Strategy, Structure, the Environment, and Financial Performance in 48 United Kingdom Companies', *Academy of Management Journal*, 23/2: 194–220.

GUEST, D. (1987), 'Human Resource Management and Industrial Relations', *Journal of Management Studies*, 24/5: 503–21.
GUEST, D. (1992), 'Right Enough to be Dangerously Wrong: An Analysis of the In Search of Excellence Phenomenon', in G. Salaman et al., *Human Resource Strategies*, London: Sage.
HALL, D. J., and SAIAS, M. A. (1980), 'Strategy Follows Structure!', *Strategic Management Journal*, 1: 149–63.
HALL, M. (1992), 'Behind the European Works Councils Directives: The European Commission's Legislative Strategy', *British Journal of Industrial Relations*, 30/4: 547–66.
HAMERMESH, R. G. (1986), *Making Strategy Work: How Senior Managers Produce Results*, New York: Wiley.
HANNAN, M., and FREEMAN, J. (1977), 'The Population Ecology of Organizations', *American Journal of Sociology*, 82: 929–64.
—— —— (1984), 'Structural Inertia and Organizational Change', *American Sociological Review*, 49: 149–64.
HEGARTY, W. H., and HOFFMAN, R. G. (1987), 'Who Influences Strategic Decisions?', *Long Range Planning*, 20/2: 76–85.
HENDRY, C., and PETTIGREW, A. (1986), 'The Practice of Strategic Human Resource Management', *Personnel Review*, 15: 3–8.
HICKSON, D. J., BUTLER, R., CRAY, D., MALLORY, G., and WILSON, D. (1986), *Top Decisions*, Oxford: Blackwell.
HICKSON, D., and MALLORY, G. (1981), 'Scope for Choice in Strategic Decision Making and the Trade Union Role', in A. Thomson, and M. Warner (eds.), *The Behavioural Sciences and Industrial Relations*, Aldershot: Gower.
HILL, C. W. L., and HOSKISSON, R. E. (1987), 'Strategy and Structure in the Multi-Product Firm', *Academy of Management Review*, 12/2: 331–41.
—— and PICKERING, J. F. (1986), 'Divisionalisation, Decentralisation and Performance of Large United Kingdom Companies', *Journal of Management Studies*, 23/1: 26–50.
HUBBARD N., and PURCELL, J. (1993), *Managing Acquisitions: The Paradox of Synergies*, annual conference of the Strategic Management Society, Sept. 1993, Chicago.
HUNT, J., and TURNER, D. (1987), 'Hidden Extras: How People Get Overlooked in Takeovers', *Personnel Management*, 19: 24–6.
Industrial Relations Review and Report (1989*a*), 'Developments in Multi-employer Bargaining: 1', *Employment Trends*, 440 (23 May).
—— (1989*b*), 'Developments in Multi-employer Bargaining: 2', *Employment Trends*, 443 (11 July).
—— (1989*c*), 'Decentralized Bargaining in Perspective', *Employment Trends*, 451 (7 Nov.).
—— (1989*d*), 'Decentralized Bargaining in Practice: 1', *Employment Trends*, 454 (19 Dec.).

References

—— (1990), 'Decentralized Bargaining in Practice: 2', *Employment Trends*, 457 (6 Feb.).
—— (1991), 'Co-ordinated Bargaining: The Debate', *Employment Trends*, 485 (5 Apr.).
—— (1992), 'Thorn lights the way to world class manufacturing', *Employment Trends*, 515 (July).
—— (1993), 'TGWU's response to lean production at Rover', *Employment Trends*, 534 (Apr.).
Institute for Employment Research (1993), *Review of the Economy and Employment 1992/3*, Coventry: University of Warwick.
JACKSON, J. H. (1986), *Organization Theory: A Macro Perspective for Management*, Englewood Cliffs, NJ: Prentice Hall.
JOHNSON, G. (1987), *Strategic Change and the Management Process*, Oxford: Blackwell.
KELLY, J. (1989), *Traditions and Socialist Politics*, London: Verso.
KESSLER, I. (1990), 'Flexibility and Comparability in Pay Determination For Professional Civil Servants', *Industrial Relations Journal*, 21/3: 194–208.
KINNIE, N. (1987), 'Bargaining within the Enterprise: Centralised or Decentralised', *Journal of Management Studies*, 24/5: 463–77.
KOCHAN, T. A., and CHALYKOFF, J. B. (1987), 'Human Resource Management and Business Life Cycles: Some Preliminary Propositions', in A. Kleingartner, and C. S. Anderson (eds.), *Human Resource Management in High Technology Firms*, Lexington, Mass.: Lexington Books.
—— KATZ, H. C., and MCKERSIE, R. B. (1986), *The Transformation of American Industrial Relations*, New York: Basic Books.
—— MCKERSIE, R. B., and CHALYKOFF, J. (1986), 'The Effects of Corporate Strategy and Workplace Innovations on Union Representation', *Industrial and Labor Relations Review*, 39/4: 487–501.
KYDD, B., and OPPENHEIM, L. (1990), 'Using Human Resource Management to Enhance Competitiveness: Lessons from Four Excellent Companies', *Human Resource Management Journal*, 29/2: 145–66.
LEGGE, K. (1978), *Power, Innovation and Problem Solving in Personnel Management*, Maidenhead: McGraw-Hill.
LEOPOLD, J., and JACKSON, M. (1990), 'Decentralization of Collective Bargaining', *Industrial Relations Journal*, 21/3: 185–208.
LLEWELLYN, D. (1981), 'Occupational Mobility and the Use of the Comparative Method', in H. Roberts (ed.), *Doing Feminist Research*, London: RKP.
LUHMAN, N. (1979), *Trust and Power*, London: Wiley.
MACINNES, J. (1985), 'Conjuring up Consultation: The Role and Extent of Joint Consultation in Post-War Private Manufacturing Industry', *British Journal of Industrial Relations*, 23/1: 93–113.

MAHONEY, J. T. (1992), 'The Adoption of the Multidivisional Form of Organization: A Contingency Model', *Journal of Management Studies*, 29/1: 49–72.

MARCHINGTON, M. (1990), 'Analysing the Links Between Product Markets and the Management of Employee Relations', *Journal of Management Studies*, 27/2: 111–32.

—— GOODMAN, J., WILKINSON, A., and ACKERS, P. (1992), 'New Developments in Employee Involvement', Research Series 2, London: Employment Department.

—— and PARKER, P. (1990), *Changing Patterns of Employee Relations*, Hemel Hempstead: Harvester Wheatsheaf.

MARGINSON, P. (1985), 'The Multidivisional Firm and Control over the Work Process', *International Journal of Industrial Organizations*.

—— ARMSTRONG, P., EDWARDS, P., PURCELL, J. and HUBBARD, N. (1993), *The Control of Industrial Relations in Large Companies: Initial Analysis of the Second Company-Level Industrial Relations Survey*, Warwick Papers in Industrial Relations, no. 45, Warwick: University of Warwick Industrial Relations Research Unit.

—— EDWARDS, P., MARTIN, R., PURCELL, J., and SISSON, K. (1987), *Beyond the Workplace: Managing Industrial Relations in the Multi-Establishment Enterprise*, Oxford: Blackwell.

—— and Purcell, J. (1986), 'The Management of Industrial Relations in UK Multi-Plant Enterprises: A New Survey Approach', *Proceedings of the 7th World Congress International Industrial Relations Association*, xvi. 143–54, Geneva: International Labour Office.

MARSH, A. (1982), *Employee Relations Policy and Decision Making*, Aldershot: Gower.

MATHEWS, J. (1991), *Colonial Mutual Life Australia: Service Quality through Self-Managing Teamwork*, UNSW Studies in Organizational Analysis and Innovation, vol. 5, University of New South Wales.

MEYER, J., and ROWAN, B. (1977), 'Institutional Organizations: Formal Structure as Myth and Ceremony', *American Journal of Sociology*, 83: 340–63.

MILLER, P. (1989), 'Strategic Industrial Relations and Human Resource Management: Distinction, Definition and Recognition', *Journal of Management Studies*, 24: 347–61.

MILLWARD, N., and STEVENS, M. (1986), *British Workplace Industrial Relations 1980–84*, Aldershot: Gower.

——, STEVENS, M., SMART, D., and HAWES, W. (1992), *Workplace Industrial Relations in Transition*, Aldershot: Dartmouth.

MINTZBERG, H. (1979), *The Structuring of Organisations*, Englewood Cliffs: Prentice Hall.

MORGAN, G. (1986), *Images of Organization*, Beverly Hills: Sage.

MOWDAY, R., PORTER, L., and STEERS, R. (1982), *Employee-Organization*

Linkages: The Psychology of Commitment, Absenteeism and Turnover, London: Academic Press.
MUELLER, F., and PURCELL, J. (1992), 'The Europeanization of Manufacturing and the Decentralization of Bargaining: Multinational Management Strategies in the European Automobile Industry', *International Journal of Human Resource Management*, 3/2: 15–35.
NEWELL, H. J. (1991), 'Field of Dreams: Evidence of "New" Employee Relations in Greenfield Sites', unpublished D.Phil thesis, Oxford University.
—— (1993), 'Exploding the Myth of Greenfield Sites', *Personnel Management*, January: 20–3.
NKOMO, S. M. (1988), 'Strategic Planning for Human Resources: Let's Get Started', *Long Range Planning*, 21/1: 66–72.
OLIAN, J. D., and RYNES, S. L. (1984), 'Organizational Staffing: Integrating Practice with Strategy', *Industrial Relations*, 23, 170– 83.
OLIVER, N., and WILKINSON, B. (1992), *The Japanization of British Industry: New Developments in the 1990s*, Oxford: Blackwell Business.
Organization of Economic Co-operation and Development (1991), *Employment Outlook 1991*, Paris: OECD.
PALMER, D., and ZHOU, X. (1993), 'Late Adoption of the Multidivisional Form by Large US Corporations: Institutional, Political and Economic Accounts', *Administrative Sciences Quarterly*, 38: 100–31.
PETERS, T., and WATERMAN R.H. (1982), *In Search of Excellence: Lessons from America's Best-Run Companies*, New York: Harper and Row.
PETTIGREW, A. M. (1975), 'Strategic Aspects of the Management of Specialist Activity', *Personnel Review*, 4: 5–13.
POOLE, M. (1980), 'Managerial Strategies and Industrial Relations', in M. Poole, and R. Mansfield (eds.), *Managerial Roles in Industrial Relations*, Aldershot: Gower.
—— MANSFIELD, B. P., and FROST, P. (1981), 'Managerial Attitudes and Behaviour in Industrial Relations: Evidence from a National Survey', *British Journal of Industrial Relations*, 20/3: 285–307.
PORTER, M. E. (1987), 'From Competitive Advantage to Corporate Strategy', *Harvard Business Review*, May: 43–59.
PRAIS, S. J. (1976), *The Evolution of Giant Firms in Britain*, Cambridge: Cambridge University Press.
PURCELL, J. (1981), *Good Industrial Relations: Theory and Practice*, London: Macmillan.
—— (1983), 'The Management of Industrial Relations in Modern Corporations: Agenda for Research', *British Journal of Industrial Relations*, 21/2: 1–16.
—— (1987), 'Mapping Management Styles in Employee Relations', *Journal of Management Studies*, 24/5: 533–48.

PURCELL, J. (1989), 'The Impact of Corporate Strategy on Human Resource Management', in J. Storey (ed.), *New Perspectives on Human Resource Management*, London and New York: Routledge.

—— (1993a), 'The End of Institutional Industrial Relations', *Political Quarterly*, 64/1: 6–23.

—— (1993b), 'The Challenge of Human Resource Management for Industrial Relations, Research and Practice', *International Journal of Human Resource Management*, 4/3: 511–27.

—— and AHLSTRAND, B. (1989), 'Corporate Strategy and the Management of Employee Relations in the Multi-divisional Company', *British Journal of Industrial Relations*, 27/3: 397–417.

—— and GRAY, A. (1986), 'Corporate Personnel Departments and the Management of Industrial Relations: Two Case Studies in Ambiguity', *Journal of Management Studies*, 23/2: 205–23.

—— MARGINSON, P., EDWARDS, P., and SISSON, K. (1987), 'The Industrial Relations Practices of Multi-Plant Foreign Owned Firms', *Industrial Relations Journal*, 18: 130–7.

—— and SISSON, K. (1983), 'Strategies and Practice in the Management of Industrial Relations', in G. Bain (ed.), (1983) *Industrial Relations in Britain: Past Trends and Future Prospects*, Oxford: Blackwell.

RAYBOULD, J. (1985), 'Ten Years of decentralisation: How Coats Patons have exploited a policy of decentralisation', *Personnel Management*, June: 40–5.

RUMELT, R. P. (1974), *Strategy, Structure and Economic Performance*, Boston: Harvard Business School Press.

—— (1982), 'Diversification Strategy and Profitability', *Strategic Management Journal*, John Wiley: 359–70.

SCHILIT, W. K., and PAINE, F. T. (1987), 'An Examination of the Underlying Dynamics of Strategic Decisions Subject to Upward Influence Activity', *Journal of Management Studies*, 24/2: 161–87.

SCHNEIDER, B. (1983), 'An Interactionist Perspective on Organizational Effectiveness', in D. Whetton, and K. S. Camerson (eds.), *Organizational Effectiveness: A Comparison of Multiple Models*, New York: Academic.

SCHULER, R. (1992), 'Strategic Human Resource Management: Linking the People with the Strategic Needs of the Business', *Organizational Dynamics*, 21/1: 18–32.

—— and Jackson, S. F. (1987), 'Organizational Strategy and Organization Level as Determinants of Human Resource Management Practices', *Human Resource Planning*, 10/3: 125–41.

SISSON, K. (1987), *The Management of Collective Bargaining*, Oxford: Blackwell.

—— and SCULLION, H. (1985), 'Putting the Corporate Personnel Department in its Place', *Personnel Management*, Dec: 36–9.

References

—— and SULLIVAN, T. (1987), 'Editorial: Management Strategy and Industrial Relations', *Journal of Management Studies*, 24/5: 427–32.

—— WADDINGTON, J., and WHITSON, C. (1992), 'The Structure of Capital in the European Community: The Size of Companies and the Implications for Industrial Relations', *Warwick Papers in Industrial Relations*, 38, Coventry: University of Warwick.

STEER, P. S., and CABLE, J. R. (1978), 'Internal Organization and Profit: An Empirical Analysis of Large UK Companies', *Journal of Industrial Economics*, 27: 13–30.

STOREY, J. (1989), 'Introduction: from Personnel Management to Human Resource Management, *New Perspectives on Human Resource Management*', London and New York: Routledge.

—— (1992), 'HRM in Action: The Truth is Out at Last', *Personnel Management*, April: 28–31.

—— (1993), 'The Take-up of Human Resource Management by Mainstream Companies: Key Lessons from Research', *International Journal of Human Resource Management*, 4/3: 529–53.

—— and BACON, N. (1993), 'Individualism and Collectivism: Into the 1990s', *International Journal of Human Resource Management*, 4/3: 665–84.

THOMASON, G. F. (1976), *A Textbook of Personnel Management*, London: IPM.

THOMPSON, R. S (1981), 'Internal Organisation and Profit: A Note', *Journal of Industrial Economics*, 30: 201–11.

THURLEY, K., and WOOD, S. (1983), 'Business Strategy and Industrial Relations Strategy', in K. Thurley, and S. Wood. (eds.), *Industrial Relations and Management Strategy*, Cambridge: Cambridge University Press.

TICHY, N. M. (1983), 'The Challenges and Context of Strategic Human Resource Management', *Human Resource Management*, 22: 45–60.

—— FOMBURN, C., and DEVANNA, M. (1982), 'Strategic Human Resource Management', *Sloan Management Review*, 22/2: 47–60.

TIMPERLEY, S. (1980), 'Organisation Strategies and Industrial Relations', *Industrial Relations Journal*, 11/5: 38–45.

TYSON, S., and FELL, A. (1986), *Evaluating the Personnel Function*, London: Hutchinson Century.

WALKER, J. W. (1981), 'The Building Blocks of Human Resource Management', *Human Resource Planning*, 4: 179–87.

—— (1988), 'Managing Human Resources in Flat, Lean and Flexible Organizations: Trends for the 1990s', *Human Resource Planning*, 11/2: 125–32.

WILLIAMSON, O. E. (1970), *Corporate Control and Business Behaviour*, Englewood Cliffs: Prentice Hall.

WILLIAMSON, O. E. and BHARGAVA, N. (1972), 'Assessing and Classifying the Internal Structure and Control Apparatus of the Modern Corporation', in *Market Structure and Corporate Behaviour*, London: Gray-Mills.

WILLMAN, P., and WINCH, G. (1985), *Innovation and Management Control: Labour Relations at BL Cars*, Cambridge: Cambridge University Press.

WILLS, T. (1984), 'Business Strategy and Human Resource Strategy', unpublished Ph.D. Dissertation: Cornell University.

WINKLER, J. T. (1974), 'The Ghost at the Bargaining Table: Directors and Industrial Relations', *British Journal of Industrial Relations*, 12/2: 191–212.

WONG, D. (1992), 'The Permanent Search for Temporary Staff: Flexible Employment Strategies or Opportunistic Labour Control?', unpublished D.Phil thesis: Oxford University.

—— (1993), 'The Permanent Search for Temporary Staff', in D. Gowler, K. Legge, and C. Clegg (eds.), *Case Studies in Organizational Behaviour and Human Resource Management*, London: Paul Chapman.

YEANDLE, D., and CLARK, J. (1989), 'Personnel Strategy for an Automated Plant', *Personnel Management*, 21: 51–5.

INDEX

ACAS 126, 135, 136
Ahlstrand, B. 63, 65, 67, 99, 130, 203
Allen, L. 22, 188
Alpander, G. 33
Andrews, K. 65, 67
appraisals 32, 116
arbitration 201
Argenti, J. 65, 176
Armstrong, P. 66
Ascher, K. 207
Associated British Ports 125
Atkinson, J. 76, 194
Australia 19, 85, 95, 128, 171, 184, 208

Bach, S. 207
Bacon, N. 181
Bamburger, P. 35
banks 141, 193
BAT 20
benchmarking 39, 108
Bhargava, N. 13
BICC 125
Botter, C. 33
Boxall, P. 32
British Aerospace 211
British Airports Authority 126
British Clothing Industry Association 125
British Leyland 205
  see also Rover
British Rail 164, 204
British Telecom 130
British Textile Employers' Association 125
BTR 21, 75
budgets 57, 60, 77, 81, 140, 156, 159, 160
  see also financial control
building societies 192
Bullock report 203
bus and coach Industry 125
Butler, R. 56
business strategy, links with HRM strategy 32–7, 41–7, 55–61, 62–74, 79–80, 129, 144, 163
  see also strategy

Cable, J. 22
Cadbury Schweppes 88, 126
Callus, R. 19
Campbell, A. 60, 66, 67, 75, 77
Capital markets:
  internal 14, 53
  international 39
Cappelli, P. 62
Career development 31, 104, 109–10, 116
CBI 110, 120–3, 158
Chalykoff, J. 58, 78
Chandler, A. 13, 24, 26, 42, 43, 50
Channon, D. 3, 21
check-off 172
Christiansen, E. 56, 60
civil service, see public service sector
Clarke, J. 201
Clegg, S. 24
Company level survey 1 56, 57, 60, 91–4, 96, 104, 163, 180, 201, 219
Company level survey 2 56, 60, 74, 91, 94, 96, 98, 114, 127, 136, 139, 163, 173, 201, 219
closed shop 171–2
Coats Viyella 125
Collective Bargaining 63, 70–2, 78, 91, 97, 118–63, 213
  choice of 143–50
  co-ordination 155–9
  corporate controls over 7, 123–4, 140, 150, 156–62
  corporate-wide 63, 138–9
  decentralized 7, 63–4, 71–72, 73, 76, 79, 97, 120–5, 139–42
  definitions 118–19
  multi employer 63, 135, 136–8, 146
  politics of 131, 150–61
  trends 120–8, 163–4
  two tier 142–3, 156
  Whitley 126, 163
  see also split bargaining: two-tier bargaining
collectivism 181–3, 211–12
  link with individualism 184–8

communication systems 92, 107, 129, 189, 192, 200
company cars 105, 108
conglomerates 15, 21
consultative committees 91, 113
  see also works council
consultancy, internal 111, 114–15
Corporate Headquarters:
  corporate personnel, *see* personnel
  culture 37, 89, 100, 107, 138, 153, 167–70
  function 15, 27
  relationship with units 54
  role 12–14, 68
corporate strategy 48
  and collective bargaining strategy 143–4
  defined 78–9
  first-order (activities and goals) 51–5, 63, 78–9
  link with HRM 32–7, 41–7, 55–61, 62–74, 79–80, 129, 144, 162
  rate of change 74
  resource implications 28, 51
  second-order (internal structure) 51–5, 63, 78–9
  significant decisions 51
  third-order (employee relations) 52–3, 63–5, 78–9
  upstream and downstream 52
cost minimization 76–7, 180–1, 186–7, 193, 212, 213
  see also styles
Crafts, N. 18
Cressey, P. 65
Critical function 87
Crompton, R. 193

Daft, R. 15
decentralization 16, 51, 53, 58, 61, 63, 70, 74, 77, 79, 84, 89, 126–7, 185–6, 210
  case studies 151–4
  corporate personnel response 71–2, 99, 132–3
  disadvantages of 144
  and importance of HRM 58
  of personnel function 76–7, 132–3, 163
  reasons for 73–4
  trade unions 131–2
Deery, S. 95
Delta Metal 125

delayering 61
Denmark 170
disputes procedure 149
diversification 16, 21, 22–3, 42, 50, 66, 72, 84, 87, 210
divestment 16, 59, 74, 167
divisionalization 16, 21, 70, 73–4, 144, 163, 210
Dock Labour Scheme 125
Donovan Report 163
Dore, R. 186
Dyer, L. 33, 34, 36, 40

Eagle Star 126
Edwards, P. 74, 76
Electrical Cable Making JIC 125
electricity supply 126, 146
Elger, T. 189
employee commitment 31, 76, 78, 80, 92, 189, 200
employee empowerment 175, 181
employers' association 94, 111, 119, 120, 129, 157
enterprise size:
  Australia 19
  European Community 19–20
  UK 17–18, 20–1
  USA 19
environment:
  changing 28–9
  external 98–9
  less negotiable 23
  turbulent 38–9
equal opportunity policies 98, 101, 102, 108, 141, 192–3
European Union single market 38
  social market 99, 111, 139, 207, 213
  Works Council 111, 139, 208

Fair Wage Resolution 131
fast food outlets 194
Federation of London Clearing Banks 125
Fell, A. 46
Ferner, A. 171
Ferranti 75
financial control 14, 60, 66, 74–7, 80, 161, 165
firms:
  foreign-owned 91–5, 116
  growth 66, 78–9, 81, 209
  high technology 189
  market power 165

privatized 38, 126, 172
professional services 189–90
size 2, 17–18, 91, 165
Fligstein, N. 24, 25
Ford 87, 119, 188
Foulkes, F. 30
Fox, A. 177–9
France 171
Franco, L. 17, 22–3
Freeman, R. 25, 199

Galbraith, J. 24
Gallaher 88, 103
GEC 66, 75, 119, 145
Germany 136, 171, 201, 208
globalization 38
Goold, M. 60, 66, 67, 75, 77
Gospel, H. 62, 129, 171
government role:
  collective bargaining structure 131
  decentralization 127
  restriction of monopoly power 165
Gowler, D. 83
Grant, R. 36, 62
Gray, A. 4, 9, 86, 97
greenfield sites 18, 31, 80, 112, 127, 172, 191, 197, 201, 206, 207
Grinyer, P. 22
Guest, D. 30–1, 32, 47, 76, 175

Hall, D. 45
Hall, M. 208
Hamermesh, R. 53, 65, 66
Hannan, M. 25
Hanson 20, 21, 66, 75, 76
health and safety 108
Hegarty, W. 55
Hendry, C. 39
Hewlett Packard 30, 47, 65, 190
Hickson, D. 2, 51, 56, 57
Hill, C. 22, 51, 53, 66, 68, 70, 71–2, 168
Hoffman, R. 55
holding companies 21–2
Honda 174
Hoskisson, R. 51–3, 66, 73
hotels and catering 126, 180, 194
Hubbard, N. 74
human resource management (HRM):
  definitions 30–2
  hard 31
  links with financial control systems 75–7
  proactive 34–5, 54

reactive 33–4, 54
soft 31, 47, 175–6, 188–91, 190, 210
and strategy 32–7, 41–7, 55–61, 62–74, 79–80, 129, 144, 163
strategy concept 29, 64, 76
see also management style, personnel management
Hunt, J. 56, 61
Hyman, R. 171

IBM 30, 47, 65, 100, 190
ICI 119, 130
Inchcape 20
individualism 130, 179–82, 211–12
  link with collectivism 184–8
industrial relations 1–2, 6, 143, 148
  adversarial 184, 187, 212
  decline of national system 170–1
  Japanese approach 174–5
  perception of new choices 174–5
  separate from personnel 185–6
insurance companies 193, 207
institutional separation 136
investment decisions 93
IPM 169–70
Italy 171
ITV companies 125

Jackson, J. 12, 24
Jackson, M. 125
Jackson, S. 33
Japan 184
Japanese companies 67, 174–5, 201
job evaluation 91, 97, 98, 126, 138–9, 154
Johnson, G. 28–9, 45, 51
joint problem solving 184, 198
joint ventures 74, 81
Jones, G. 193

Katz, H. 2, 37, 40–1, 63, 66, 77, 128
keizen (continuous improvement) 175
Kelly, J. 172
Kessler, I. 126
Kinnie, N. 9, 74, 124, 162
Kochan, T. 2, 37, 40, 41, 58, 63, 66, 77–8, 128
Konosuke Matsushita 175
Kydd, B. 33, 36

labour:
  as a commodity 180, 182, 212
  costs 76, 77, 130

labour (cont.):
  exploitation 186–7, 194–5, 213
  legislation 98–9, 129, 131, 172
  male/female split 192–3
  as a resource 180, 181–2, 188, 211
  unskilled 186–7, 194
labour markets:
  configuration 143
  corporate power 14
  deregulation 39
  external 148
  internal 64, 72–4, 148, 168, 180, 186, 211
Legal and General 126
Legge, K. 80, 83, 90
Leopold, J. 125
line management 150
  industrial relations 129–30
  and personnel department 82–3
  transformational leadership 214
Llewellyn, D. 193
lobbying 110–11
local authorities, *see* public service sector
Lucas 126
Luhman, N. 188

MacInnes, J. 176, 199
McKersie, R. 2, 37, 40–1, 58, 63, 66, 77, 128
Mahoney, J. 24
mail order business 194–5
Mallory, G. 2, 55
management accounting 66, 161
management development and succession 72–3, 85, 103–4, 109–10, 116, 170, 191
management style 65–7, 73, 79, 100, 165–215
  adversarial 184, 187–8, 203–4
  bargaining constitutional 196–7
  collectivism 138, 212
  cooperative 184
  cost minimization 180, 186, 207–8, 212
  employee development 180, 181, 207, 211
  matrix 176–201
  modern paternalist 197–9
  paternalism 180–1, 191–3, 212
  sophisticated consultative 199–201, 207, 214

  sophisticated human relations 188–91
  traditional 193–6, 213
  unitary and pluralist frames of reference 177–9
  unstable styles 186–7, 202–4, 212–13
manufacturing, decline 18–19
Marchington, M. 37, 38, 181–2
Marginson, P. 56, 58–9, 60, 74, 77, 94, 96, 97, 104, 127, 128, 141, 161, 173, 180
Marks and Spencer 47, 65, 87
Marsh, A. 56, 94, 114, 123
Massey Ferguson Perkins Engines 126
Mathews, J. 208
Meager, N. 76
Meakin, R. 202
Medoff, J. 199
meat and dairy industry 126
mergers and acquisitions 18, 21, 38, 56, 57, 58, 66, 67, 74, 75, 77, 81, 94, 108–9, 142, 210
Metal Box 126
Meyer, J. 26
Midland Bank 125
Miller, P. 55
Millward, N. 56, 93, 94, 96, 124, 127, 172
mixed bargaining 158
Mintzberg, H. 75, 76
Morgan, G. 14
Mueller, F. 200
multi-divisional company (m-form):
  advantages 23–4
  definition 11–16
  dominance of 2 97–8
  growth of 16–23, 97
  theories of 24–6
  types of economies 52–4
  variations 15–16
  see also decentralization; diversification; divestment; divisionalization
multi-employer bargaining 119, 120, 126, 149, 163
  decline 125, 170–1
  pros and cons 135–8
multinational expansion 109
Multiple Food Retailers Association 125

National Health Service 126, 142
  see also public service sector
National Power 149

Netherlands 171, 208
new industrial relations 70
New Zealand 128, 171
Newell, H. 31, 207
newspaper distribution and printing 125
Nissan 174, 201
Nkomo, S. 36, 56, 60
Norway 171
Nuclear Electric 149
numerical flexibility 180, 186–7, 194

Olian, J. 33
Oliver, N. 175
Oppenheim, C. 33, 36
organizational development 33, 111

P&O 20
paint and varnish industry 125
Paine, F. 71
Palmer, D. 24
Parcel Force 144
Parker, P. 181–2
part-time workers 80, 194–5
paternalism 181, 190–2, 211, 212
  modern 196–9
  movement away from 206–7
pension funds 112–13, 134
pensions 112–13
performance control 13–14, 23, 27, 47, 161–2
performance related pay 31, 130, 143, 187, 194, 200
performance targets 75–7, 80, 166
personnel management 30, 38, 45–6, 55–6, 61, 63, 76, 80, 169–70, 205
  in corporate office 56, 60, 63–4, 67–70, 71, 73, 79, 82–117, 159–61
  decline of 85, 94, 96, 104–5
  and firm structure 87–91
  guidelines policies 101–2
  information systems 111
  marketing of services 114–15
  optimum size 87–100
  personnel directors 56–7, 83–4, 94–5, 96, 169
  policy committees 94–5, 103–4, 116
  response to decentralization 71–2, 99, 132–3
  vested interests 133, 151–2
  see also human resource management
Peters, T. 35, 55, 107

Pettigrew, A. 39, 46
Philips, B. 35
Pickering, J. 22, 51, 68–70, 72, 168
Pilkington 126
Pirelli General 125, 201
political power 14
political theory 25, 45, 49
politics 37, 48
Poole, M. 176, 179
population ecology theory 25
Porter, M. 42, 76
portfolio planning 53–4, 66, 76, 78, 166
Postal Counters 144
power 14, 45–6, 58
  theory of 90
Powergen 149
Prais, S. 21
privatization 38, 126, 172–3, 207
productivity 144
productivity bargaining 130, 139, 203
profit centres 71, 78, 97, 133, 145, 151
  decentralization 89
  profit sharing 92
  psychometric testing 32, 110, 188
  and strategy responsibility 12–13, 42
public service sector 16, 23, 26, 38, 51, 71, 116, 125, 126, 130, 142, 163, 171
Purcell 3, 4, 9, 32, 42, 63, 65, 66, 86, 92–3, 97–8, 126, 170, 181, 187, 200

quality 31
Quality circles 76, 77

Racal 126
Raybould, J. 115
recession, effect of 47
research programme 4–8
Rover Motor Company 174, 201–2, 204, 211, 216–18
Rowan, B. 26
Royal Insurance 126
Royal Mail 142, 144, 204–5
Rumelt, R. 16–17, 22
Rynes, S. 33

Saias, M. 45
Schedule, II. 131
Schilit, W. 71
Schneider, B. 33
Schuler, R. 33–4, 37
scientific management 174

Scullion, H. 46, 67, 86–8, 91, 100, 103–4
security services 103
share option schemes 32
share ownership 92
share prices 77
shareholders, institutional 39
shop stewards 185
combine committees 139, 142
short termism 75–7, 80, 215
Singh, H. 63
Sisson, K. 20, 46, 55, 65, 67, 86, 87–8, 91, 97, 100, 103
Social chapter, *see* European Union
South Africa 186
split bargaining 119, 142
stakeholders 67
Steel corporation 125
Steer, P. 22
Stephenson, G. 188
Stevens, M. 56, 93, 96, 124, 127
Storey, J. 31, 32, 182, 205, 206
strategic business unit (SBU) 15, 165
strategies and strategic choice 37–42, 165–6, 171
 in collective bargaining structure 134–6
 decisions 28, 45, 71
 definitions 27–9, 49
 distinction between corporate and business 166
 personnel organization 88–91
 resource view 62
 three levels of 42–7, 51–5, 61, 62–3, 74, 78–9, 164, 167
 upstream and downstream 42–3, 49
 see also business strategy; corporate strategy; human resource management
subcontracting 207, 213
Sullivan, T. 65
Sweden 137, 171, 208
Synergies 66, 78, 110

take-over bids 77
Tandy Corporation 100
Tarmac 75
Taylor, F. W. 174
technology 38–9
temporary workers 180, 194, 195
Texas Instruments 30
Thailand 196
Thomason, G. 40

Thompson, R. 22
Thorn Lighting 202, 206
Thurley, K. 37, 41
Tichy, N. 33, 35
Timperley, S. 37, 41
Tootal 125
Total Quality Management 32, 77, 107, 175, 189, 191, 200, 213
Toyota 174, 202
Trade Union Reform and Employment Rights Act 172
Trade unions 82, 155, 171–2, 176, 183, 186, 197, 201, 204–5, 210
 adaptation to HRM 206
 decentralization 127, 131–2, 160–1
 management attitude to 187–8, 191–3, 208, 213
 membership 171–3, 204
 national officers 71, 139, 146, 152, 155, 158
 non-union firms 58, 78, 100, 189–90, 208
 power 73–4, 98, 129, 171
 recognition 58, 60, 98, 163, 172, 204–5
 training 97, 112, 114–15
 see also management development
transaction cost 141, 144
 theory 24–5
Turner, D. 56
two-tier bargaining 119, 125, 130, 142–3, 156
Tyson, S. 46

uncertainty 28, 39, 49, 51, 53
unemployment, long term 171
Unipart 204
United Biscuits 126
USA 16–17, 19, 20, 30, 56, 60, 63, 80, 128, 171, 184, 196, 201, 208

vertical integration 52–3

Walker, J. 33
Waterman, R. 35, 55, 107
water industry 125
Whitbread 89
Wilkinson, B. 175
Williams Holdings 21
Williamson, O. 13, 24, 50, 67
Willman, P. 205
Wills, T. 35
Winch, G. 205

Winkler, J. 2, 55, 169
Wong, D. 194–5
Woods, S. 37, 41
works council 183, 201, 208, 210
  see also European Union

world trade 137, 165

Yeandle, D. 201

Zhou, X. 24